Quasiconvex Optimization and Location Theory

Applied Optimization

Volume 9

Series Editors:

Panos M. Pardalos
University of Florida, U.S.A.

Donald Hearn
University of Florida, U.S.A.

Quasiconvex Optimization and Location Theory

by

Jaoquim António dos Santos Gromicho
ORTEC Consultants bv,
Gouda, The Netherlands

KLUWER ACADEMIC PUBLISHERS
DORDRECHT / BOSTON / LONDON

A C.I.P. Catalogue record for this book is available from the Library of Congress.

ISBN-13: 978-1-4613-3328-9 e-ISBN-13: 978-1-4613-3326-5
DOI: 10.1007/978-1-4613-3326-5

Published by Kluwer Academic Publishers,
P.O. Box 17, 3300 AA Dordrecht, The Netherlands.

Sold and distributed in the U.S.A. and Canada
by Kluwer Academic Publishers,
101 Philip Drive, Norwell, MA 02061, U.S.A.

In all other countries, sold and distributed
by Kluwer Academic Publishers,
P.O. Box 322, 3300 AH Dordrecht, The Netherlands.

Printed on acid-free paper

Para os meus Pais,
que, de entre todos aqueles que acham o meu trabalho incompreensível,
continuam a ser aqueles que dele mais se orgulham.

Contents

List of Figures

List of Algorithms

List of Tables

Acknowledgments

This book is a slightly enhanced version of a previously published manuscript, my Ph.D. thesis (Ref. [50]). Most of the effort needed to write it has, in fact, taken place during my Ph.D. research and as such I should start by thanking again all the people that were acknowledged in the original manuscript. There is no better way to do this than quoting the original text:

> Since every story starts at the beginning, I must start by thanking Hans Frenk and Martine Labbé for having come to Lisbon in the Spring of 1990 and for having suggested the possibility of undergoing a Ph.D. program at the Erasmus University Rotterdam. From a certain Spring day, walking on a beach while listening to Hans with his characteristic enthusiasm talking about future possibilities, to a certain Winter day (January 8, 1991), marking the arrival at Rotterdam, everything seems now in my memory to have been surprisingly simple. Two entities deserve equal credits for having made this possible, the *Departamento de Estatística e Investigação Operacional* of the Lisbon University for giving me a leave of absence and the *Junta Nacional de Investigação Científica e Tecnológica* (J.N.I.C.T.) for giving me the grant BD/631/90-RM. Moreover, I would also like to acknowledge the Tinbergen Institute Rotterdam for their financial support during the fourth year of my stay.
>
> Now that I look back, it was quite a jump in the dark to accept the challenge of coming here. The enthusiasm of Hans, the name of Alexander Rinnooy Kan, well-known to me as a student in Operations Research, the company of Ana Isabel on embarking together in this project and the support of my parents are highly responsible for having done so. I feel very happy that it was possible to arrange everything. My life in Rotterdam has been extremely fulfilling and enjoyable both from a scientific and a personal point of view.

Scientifically, it has been a challenge since the very beginning. Being assigned to work with Hans, I soon discovered that the cheerful and good-humored Hans who talked me into this while walking on the beach, is the most committed supervisor that I have ever met. His tenacity and the experienced supervision of Alexander gave me a very good start. Shuzhong, who was finishing his dissertation in those days, helped me a lot catching up with the frenetic rhythm that it takes to work under the supervision of Hans. Having met Frank Plastria for the first time in Tenerife during a congress which I attended, among others, under the support of J.N.I.C.T., proved to be very fruitful.

The entire staff of the Mabes group is highly acknowledged for a nice working environment. Within the group, the Portuguese community deserves special appraisal. Ana Isabel and Teresa, together with at least a once-per-week letter and frequent phone calls of my parents, greatly helped making the experience worthwhile. The amount of nice people that I have met during my stay here is too large to mention them all. Hence, my excuses in advance for all the involuntary omissions. Many people read and commented, some of them several times, parts or the whole of my thesis. For their valuable comments, special thanks go to Hans, Alexander, Shuzhong, Frank Plastria, Ana Isabel, Teresa and Marcel Kleijn. Albert and Nanda are acknowledged for having brought some sense into my *Samenvatting.* Nevertheless, any mistakes that may be found in the manuscript are my own responsibility.

Finally, the support of Ana and Auke in moments of stress meant a lot to me. I am counting with this (hopefully by then!) pair of Doctors to back me up on the day of my defense! Last but not least, the continuous support and encouragement of my parents, although from far away, and the recent support of Marlies, are highly acknowledged.

Destiny plays sometimes funny tricks. I wrote the above text in Rotterdam, in the beginning of 1995, while finishing the original manuscript and preparing for defence. After acquiring my title, I went back to Portugal. One year ago, just before Easter 1996, I went to Hungary to attend a conference at Màtrahàza, on top of the beautiful Màtra mountain. I had a good time there, both from scientific and personal points of view. Among the many nice people I met, two would be decisive to make this book appear: John Martindale and Panos Pardalos. Their encouragement to rewrite this book to meet Kluwer's standards are highly acknowledged. Their patience too. Specially John's.

On my way back to Portugal I spent Easter holidays in The Netherlands. During those holidays I received a tempting offer to join a great consulting company in Operations Research, ORTEC. I ended up accepting, and that is the reason why, ironically, this book has been finished, and the acknowledgements written, once again in Rotterdam. During this new stay in The Netherlands two persons from the previously quoted acknowledgements demonstrated that they deserve to be mentioned again: Hans Frenk for his continuous scientific support and Marlies for her patience.

Further, I should acknowledge Emilio Carrizosa, Frank Plastria, Mikhail Andramonov, Siegfried Schaible, Roland Freund and Igor Konnov for their useful comments, suggestions and contributions sent as an answer to a requested I posted to the OPTNET searching for recent references in the field.

Also ORTEC, and most notably the EuroPlanner team, is acknowledged for putting up with my foggy mornings each time it would take some cups of coffee to recover from a night of "book work". The Logistics department, in the person of Marjon Gredt, is highly acknowledged for allowing me the use of the printer for this matter.

Finally, the continuous encouragement from my dear parents, whatever turns and bends I decide to bring into my life, cannot be forgotten. *Muito obrigado!*

Rotterdam, March 25, 1997
Joaquim António dos Santos Gromicho

I Introduction

A crucial step in making decisions of a quantitative nature requires the solution of optimization problems. Such optimization problems can be described as minimizing or maximizing an *objective function* subject to a family of *constraints*. The choice of which functions to use as an objective function and as constraints depends on the *modeling* of the problem. However, modeling a problem is only useful in practice if there exist methods to solve the proposed model. This justifies the enormous popularity of *linear programs*, i.e. optimization problems where the objective function and the constraints are linear functions. In fact, a practical method, the popular simplex method, exists since the late forties to solve these problems. In spite of this, there are many real life situations which cannot be appropriately modeled using only linear functions and therefore the field of *nonlinear programming* has received increasing attention from many researchers.

Nowadays, convex analysis, which started in a modern sense with the well-known book of Rockafellar (Ref. [89]), is recognized as the main analytical tool for analyzing nonlinear optimization programs. This led to the creation of a new field called *convex programming*. Clearly, convex programs are more flexible from a modeling point of view than linear programs and form an important subclass within nonlinear programming. The books of Hiriart-Urruty and Lemaréchal (Ref. [52, 53]) serve as excellent references for results on convex programs. A short overview without proofs of some results in convex analysis is given in Chapter 2 of this book.

One of the most important characteristics of convex programs is that differentiability of the objective or constraint functions is not required. In fact, *gradients* are replaced by *subgradients*. Shor (Ref. [100]) emphasizes this relation by using the term *non-differentiable optimization* as a synonym for convex programming. Another example is provided by the recent work in Ref. [88].

The vast majority of methods solving convex programs depends on the construction of a hyperplane separating at the domain level a given point from a set of points having a lower function value than this point. The success of these methods for convex programming comes from the fact that subgradients of convex functions define instances of such separating hyperplanes, as the analysis in Chapter 3 shows. However, Chapter 5 shows that such a separating hyperplane can also be constructed for the more general class of *quasiconvex functions*. This permits the solution of *quasiconvex programs*, i.e. the minimization of a quasiconvex objective function in presence of convex constraints.

The *ellipsoid method* is used as an illustration of the use of separation hyperplanes in convex and quasiconvex programming. This very powerful method evolves by defining a succession of progressively smaller ellipsoids (hence its name!) all of them including an optimal point. Each ellipsoid is obtained from the previous one by means of a separating hyperplane.

Chapter 3 and Chapter 5 define and analyze versions of the ellipsoid method suitable to solve convex and quasiconvex problems. The analysis exploits properties of these problems and in both cases convergence is established by very simple means. However, despite their simplicity, these are strong results which seem to be new.

Convex programming is well established as the "easy" part of nonlinear programming. That is due to the availability of very strong properties, such as necessary and sufficient optimality conditions (Ref. [11, 52, 89]) strong duality (Ref. [11, 52, 90]) and efficient algorithms (Ref. [11, 53, 75, 100]).

However, all these properties are valid in slightly more general cases, as the above mentioned books also refer. The study of the most general settings in which desirable properties of convex programming still hold gave rise to so-called weak forms of convexity, or as they are also known, general forms of convexity (Ref. [85]). Among others, Schaible has been a true paladin of this study (Ref. [7, 15, 61, 93]) motivating increasing interest and research efforts aiming at the clarification of the exact role played by each aspect of convexity.

A related filed that has received a lot of attention in the past years is the field of *fractional programming* (Ref. [4, 10, 15, 17, 20]). This can be simply put as the study of optimization problems where the objective function is the ratio of two functions. Recently Freund and Jarre (Ref. [42, 43]) applied the interior point methodology to this field and developed interior point methods for (differentiable) fractional programs of which the objective is given by the ratio of a convex by a positive (over a convex domain) concave function.

As observed by Sniedovich (Ref. [102, 103]) most of the properties of fractional programs could be found in other programs, given that the objective function could be written as a particular *composition* of functions. He called this new field *C-programming*, standing for *composite concave programming.* In his seminal book on *dynamic programming* (Ref. [104]), Sniedovich shows how the study of such compositions can help tackling non-separable dynamic programs that otherwise would defeat solution. Barros and Frenk (Ref. [9]) developed a cutting plane algorithm capable of optimizing C-programs. More recently, this algorithm has been used by Carrizosa and Plastria to solve a global optimization problem in facility location (Ref. [16]).

The distinction between *global optimization* problems (Ref. [54]) and generalized convex problems can sometimes be hard to establish. That is exactly the reason why so much effort has been placed into finding an exhaustive classification of the different weak forms of convexity, establishing a new definition just to satisfy some desirable property in the most general way possible. This book does not aim at all the subtleties of the different generalizations of convexity, but concentrates on the most general of them all, quasiconvex programming. Chapter 5 shows clearly where the real difficulties appear.

We do not impose differentiability of these functions but aim at describing separators of the (convex) lower level set of such functions when the gradient does not exist. In fact, we try to characterize the corresponding generalizations of subgradients. Plastria (Ref. [80]) introduced the concept of *lower subdifferentiability.*

This concept holds for an important subset of the quasiconvex functions, the appropriately called *lower subdifferentiable functions*, and is mostly a property of (lower parts of) the epigraph of the functions. We look just at the domain of the function, looking for elements of the normal cone to the (convex) lower level sets.

Andramonov called our attention to the fact that much research has been published in Russian (and as such, unfortunately unavailable to us) in fields that highly intersect those focused in this book. Most of it in the University of Kazan. In particular, Andramonov (Ref. [1–3]) obtained conditions of quasiconvexity of so-called *marginal functions* and constructed a method of conical projections for quasiconvex programming. It can be considered as an interior point method, transforming the initial problem into a problem of unconstrained minimization. In the same University, Zabotin (Ref. [116–119]) studies optimality conditions for quasiconvex programs through normal cones to the level sets and proves convergence of generalized gradient descent methods. His results seem to relate in some extent to results in this book. Also from Kazan University, Konnov wrote a book (Ref. [62]) covering aspects of quasiconvex optimization. Another reference to Konnov's work (fortunately in English!) is given by reference Ref. [63].

Since the research reported in the present book has been started by the need to solve general convex and quasiconvex location problems, general models for this problems with convex and quasiconvex objective functions are presented in Chapter 4 and Chapter 6 respectively. The solution methods developed in Chapter 3 and Chapter 5 are applied to this problems and some computational results are given.

Although these nonlinear real life situations can be found in many fields we restrict ourselves in this book to the field of continuous location. Historically most of the classical continuous location models gave rise to linear programs by taking an appropriate distance measure. However, as techniques to solve nonlinear programs emerged, also more realistic nonlinear continuous location problems appeared in the literature. A good survey on the impact of the new techniques on the proposed models is given by Plastria in Ref. [83].

Finally, in Chapter 7 we present some conclusions and directions of future research.

II Elements of Convexity

This chapter reviews and introduces some concepts which are important throughout this book. It starts in Section 2.1 by introducing some notational conventions. The chapter proceeds by discussing some elementary properties of convex sets in Section 2.2 and of convex functions in Section 2.3. Finally it ends with Section 2.4 where some concepts introduced in the previous sections are generalized.

2.1. Generalities

The first occurrence of a term is emphasized in this book by writing the term in *italics* and can be seen in many cases as a formal definition.

Our working space is the *s-dimensional Euclidean space*, $\mathbf{R}^s$. Elements of $\mathbf{R}^s$ are sometimes called *vectors*, sometimes *points*, but always represented by bold lower case letters, e.g. $\boldsymbol{x} \in \mathbf{R}^s$. In pictures they are represented by *arrows*, $\longrightarrow$, if their vectorial properties are relevant or by *dots*, $\bullet$, if they should be interpreted as points. The *null vector*, i.e. the vector with all components equal to zero, is represented by $\mathbf{0}$. *Matrices* are represented by bold capitals, e.g. $\boldsymbol{A} \in \mathbf{R}^{s\times s}$. By adding the superscript $\top$ as in $\boldsymbol{A}^\top$ we denote the *transpose* of $\boldsymbol{A}$. When vectors are represented by listing their coordinates it is understood that

$$\boldsymbol{x} = (x_1, x_2, \ldots, x_s) = \begin{bmatrix} x_1 \\ x_2 \\ \vdots \\ x_s \end{bmatrix}.$$

A vector is always a *column matrix* and hence one can write $\boldsymbol{x}^\top \boldsymbol{A}\boldsymbol{y}$. We denote by $\boldsymbol{x}^\top\boldsymbol{y}$ the usual *inner product* of $\boldsymbol{x}, \boldsymbol{y} \in \mathbf{R}^s$ and by $\|\boldsymbol{x}\|_2 := \sqrt{\boldsymbol{x}^\top\boldsymbol{x}}$ the *Euclidean norm.* When we write $\boldsymbol{a} \leq \boldsymbol{b}$ with $\boldsymbol{a}, \boldsymbol{b} \in \mathbf{R}^s$ we mean $a_i \leq b_i$ for every $1 \leq i \leq s$.

Sets are represented by calligraphic capitals, e.g. $\mathcal{A}$, when they are subsets of $\mathbf{R}^s$ and by slanted capitals, e.g. $I := \{1, 2, \ldots, s\}$, when they are sets of *indices.* $\mathbb{N}$ denotes the set of positive integers, i.e. $\mathbb{N} := \{1, 2, \ldots, n, n+1, \ldots\}$.

Real numbers are sometimes called *scalars* and when represented by literals these are slanted symbols, e.g. $\alpha, x \in \mathbf{R}$ and $\Delta, L > 0$. The set $\mathbf{R} \cup \{-\infty, +\infty\} = [-\infty, +\infty]$ is denoted by $\overline{\mathbf{R}}$. *Intervals of real numbers* are represented as

$$
\begin{aligned}
[\mathrm{a,b}] &:= \{x \in \mathbf{R} : a \le x \le b\}, &]\mathrm{a,b}] &:= \{x \in \mathbf{R} : a < x \le b\}, \\
[\mathrm{a,b}[&:= \{x \in \mathbf{R} : a \le x < b\} \quad \text{and} &]\mathrm{a,b}[&:= \{x \in \mathbf{R} : a < x < b\}.
\end{aligned}
$$

Line segments in $\mathbf{R}^s$ are denoted (analytically on the left, in pictures on the right) by

$$
\begin{aligned}
[\boldsymbol{a},\boldsymbol{b}] &:= \{(1-\lambda)\boldsymbol{a} + \lambda\boldsymbol{b} : \lambda \in [0,1]\} \\
]\boldsymbol{a},\boldsymbol{b}] &:= \{(1-\lambda)\boldsymbol{a} + \lambda\boldsymbol{b} : \lambda \in]0,1]\} \\
[\boldsymbol{a},\boldsymbol{b}[&:= \{(1-\lambda)\boldsymbol{a} + \lambda\boldsymbol{b} : \lambda \in [0,1[\} \\
]\boldsymbol{a},\boldsymbol{b}[&:= \{(1-\lambda)\boldsymbol{a} + \lambda\boldsymbol{b} : \lambda \in]0,1[\}
\end{aligned}
$$

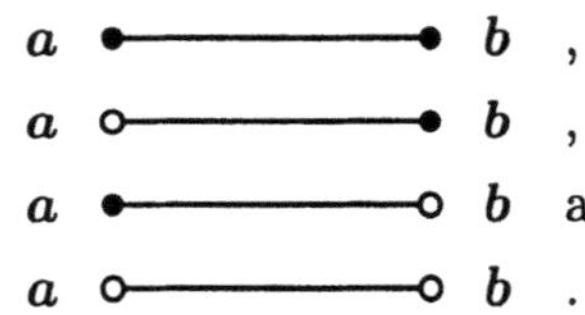

In pictures, boundaries of *open sets* are represented by *dashed lines* - - - - - - - - - - while boundaries of *closed sets* are represented by *full lines* ———————— . The concepts of open and closed sets are introduced in the next section. If it is important to stress the membership of particular points then closed (or open) small balls are used to denote points that are part (or not) of the set. An illustration is provided by the above pictorial representation of line segments.

When a *proof* of a result is given it follows the result itself and is started by the bold word **Proof:** and ended by the symbol □.

Each picture in a captioned figure is displayed inside a shadowed box while algorithms with a caption are displayed inside a thick oval box. Procedures, or other algorithms detailed elsewhere, appearing within displayed algorithms are framed by an oval box and referred to as *framed statements.*

This chapter proceeds now by introducing the basic concepts and presenting without proofs some results which are used throughout this book.

2.2. Convex sets

A convex set is a set that contains the line segment connecting every pair of points included in it. A formal definition is the following.

Definition 2.2.1 (Ref. [52, 89]). *The set $C \subset \mathbf{R}^s$ is convex if $[x, y] \subseteq C$ for each $x, y \in C$.*

Figure 2.1 shows two sets: the one on the left is convex and the one on the right is not. The set on the right is not convex since the represented line segment connecting a pair of points of the set has its dotted part outside the set.

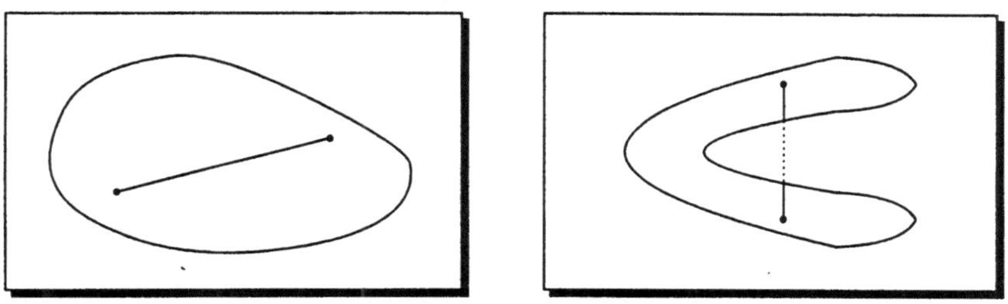

Figure 2.1: A convex and a nonconvex set

The convexity of sets is preserved by certain operations. As an example we mention that the intersection of convex sets is a convex set (Proposition III.1.2.1 of Ref. [52]). Moreover, if $\lambda\mathcal{S} := \{\lambda x : x \in \mathcal{S}\}$ and $\mathcal{S}_1 + \mathcal{S}_2 := \{x_1 + x_2 : x_1 \in \mathcal{S}_1 \text{ and } x_2 \in \mathcal{S}_2\}$ then by Proposition III.1.2.4 of Ref. [52] these sets are convex if $\mathcal{S}$, $\mathcal{S}_1$ and $\mathcal{S}_2$ are convex.

In order to introduce topological concepts of (convex) sets we denote the *open Euclidean unit ball* by $\mathcal{B} := \{x \in \mathbf{R}^s : \|x\|_2 < 1\}$.

Given a general subset $\mathcal{S}$ of $\mathbf{R}^s$, a point $x \in \mathcal{S}$ is called an *interior point* of $\mathcal{S}$ if there exists some $\delta > 0$ such that $x + \delta\mathcal{B} \subseteq \mathcal{S}$. The set of all interior points of $\mathcal{S}$ is the *interior* of $\mathcal{S}$ and is denoted by $\text{int}(\mathcal{S})$. A point $x \in \mathbf{R}^s$ with the property that $(x + \delta\mathcal{B}) \cap \mathcal{S} \neq \emptyset$ and $(x + \delta\mathcal{B}) \setminus \mathcal{S} \neq \emptyset$ for every $\delta > 0$ is called a *boundary point* of $\mathcal{S}$ and the set of all such points is denoted by $\text{bd}(\mathcal{S})$. The *closure* of $\mathcal{S}$ is $\text{cl}(\mathcal{S}) := \mathcal{S} \cup \text{bd}(\mathcal{S})$. The set $\mathcal{S}$ is called *open* if it equals its interior and *closed* if it equals its closure. Also the set $\mathcal{S}$ is said to be *bounded* if there exists some $M > 0$ such that $\mathcal{S} \subseteq M\mathcal{B}$. A bounded and closed subset of $\mathbf{R}^s$ is called *compact.*

Returning now to convex sets, *hyperplanes* are very important instances of them. A hyperplane is defined by a vector $\boldsymbol{a} \in \mathbf{R}^s$ and by a scalar $b \in \mathbf{R}$ as $\mathcal{H}^{=}(\boldsymbol{a}, b) := \{\boldsymbol{x} \in \mathbf{R}^s : \boldsymbol{a}^\mathsf{T}\boldsymbol{x} = b\}$. Each hyperplane defines four halfspaces

$$\begin{aligned}
\mathcal{H}^{\leq}(\boldsymbol{a}, b) &:= \{\boldsymbol{x} \in \mathbf{R}^s : \boldsymbol{a}^\mathsf{T}\boldsymbol{x} \leq b\},\\
\mathcal{H}^{<}(\boldsymbol{a}, b) &:= \{\boldsymbol{x} \in \mathbf{R}^s : \boldsymbol{a}^\mathsf{T}\boldsymbol{x} < b\},\\
\mathcal{H}^{\geq}(\boldsymbol{a}, b) &:= \{\boldsymbol{x} \in \mathbf{R}^s : \boldsymbol{a}^\mathsf{T}\boldsymbol{x} \geq b\} \text{ and}\\
\mathcal{H}^{>}(\boldsymbol{a}, b) &:= \{\boldsymbol{x} \in \mathbf{R}^s : \boldsymbol{a}^\mathsf{T}\boldsymbol{x} > b\}.
\end{aligned}$$

By pure convention (since in general dimensions the notions of "top" and "bottom" are ambiguous), $\mathcal{H}^{\leq}(\boldsymbol{a}, b)$ is called a *closed lower halfspace* and $\mathcal{H}^{\geq}(\boldsymbol{a}, b)$ a *closed upper halfspace.* This convention is supported by imagining that the vector $\boldsymbol{a}$, the so-called *normal* vector to the hyperplane, "points up" and observing that the lower halfspace is the one on the opposite side of this hyperplane. Finally, $\mathcal{H}^{<}(\boldsymbol{a}, b)$ and $\mathcal{H}^{>}(\boldsymbol{a}, b)$, are the *open* versions of the previous sets. From this point on, unless stated otherwise, "halfspace" always refers to "closed lower halfspace". Also, it is easy to verify that halfspaces are convex sets.

An important instance of convex sets defined as intersection of halfspaces is given by the following definition.

Definition 2.2.2 (Ref. [52, 89]). *A closed convex polyhedron $\mathcal{P}$ is the intersection of finitely many halfspaces, i.e. $\mathcal{P} := \bigcap_{1 \leq j \leq m} \mathcal{H}^{\leq}(\boldsymbol{a}_j, b_j) = \{\boldsymbol{x} \in \mathbf{R}^s : \boldsymbol{a}_j^\mathsf{T}\boldsymbol{x} \leq b_j,\ j = 1, \ldots, m\}$. If $\boldsymbol{A} \subseteq \mathbf{R}^{m \times s}$ is the matrix whose rows are the vectors $\boldsymbol{a}_j \in \mathbf{R}^s$ and $\boldsymbol{b} \in \mathbf{R}^m$ is the vector with components b_j, $1 \leq j \leq m$, then the matrix notation $\mathcal{P} = \{\boldsymbol{x} \in \mathbf{R}^s : \boldsymbol{A}\boldsymbol{x} \leq \boldsymbol{b}\}$ is often used. If the polyhedron is bounded it is called a polytope.*

An example is provided by Figure 2.2 where six lower halfspaces and their normal vectors $\boldsymbol{a}_j$, $1 \leq j \leq 6$, are represented.

Although convex polyhedrons are very special closed convex sets they play an important role among convex sets.

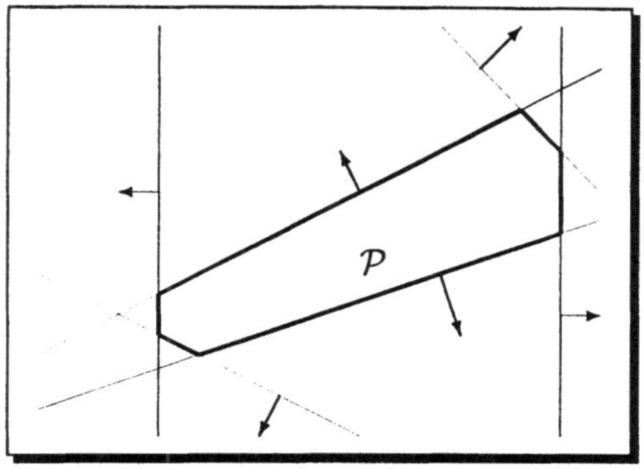

Figure 2.2: A polytope resulting from the intersection of six halfspaces

As a consequence of a well-known separation result (Corollary 2.2.1 in Subsection 2.2.3) it can be shown that every closed convex set C is the intersection of (maybe infinite in number) halfspaces containing C (Theorem III.4.2.4 of Ref. [52]). Hence, each closed convex set can be approximated as well as desired by a convex polyhedron. A similar result holds for an open convex set if open halfspaces are allowed (Lemma III.4.2.1 of Ref. [52]).

Another important instance of convex sets is now defined.

Definition 2.2.3 (Ref. [32]). *A set $C \subseteq \mathbf{R}^s$ is called evenly convex if C is the intersection of open halfspaces, i.e.* $C := \bigcap_{j \in J} \mathcal{H}^{<}(a_j, b_j) = \{x \in \mathbf{R}^s : a_j^\top x < b_j,\ j \in J\}$.

Clearly, since every open convex set is the intersection of open halfspaces it is evenly convex. By observing that closed halfspaces can be obtained by intersecting open halfspaces, e.g. $\mathcal{H}^{\leq}(a, b) = \bigcap_{t>0} \mathcal{H}^{<}(a, b + 1/t)$, one concludes that every closed convex set is also evenly convex. However, convex sets exist that are not evenly convex. In Figure 2.3 we illustrate how such a set can be constructed. Start with a rectangle as in the first picture. This set is evenly convex, since it is closed and convex. Now remove one of its edges. In the second picture the top edge is chosen. This set is still evenly convex since it results from the previous one by intersecting it with a new open halfspace, namely the one whose hyperplane includes the missing edge. Finally, add one of the erased vertices, as done in the last picture where the top right vertex is added. This last set is still convex, but it cannot be obtained as the intersection of open halfspaces. Hence, it is not evenly convex.

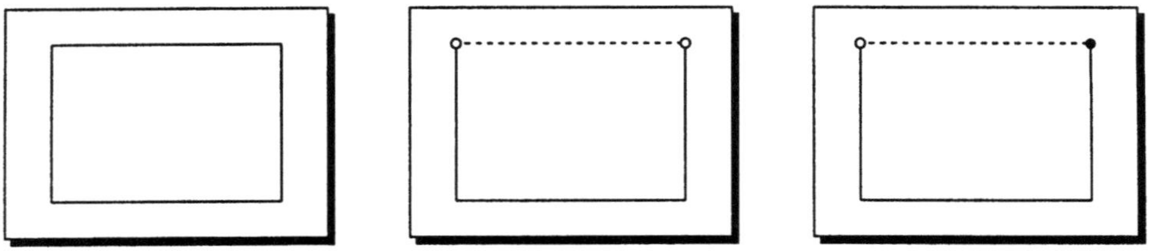

Figure 2.3: Constructing a convex set that is not evenly convex

Evenly convexity differs from convexity only at the *relative boundary.* This term, and others, are introduced in Subsection 2.2.2. The next subsection is dedicated to *hulls.*

2.2.1. Hulls

Let a set $\{\boldsymbol{x}_1, \boldsymbol{x}_2, \ldots, \boldsymbol{x}_n\}$ of points be given. The point $\boldsymbol{x}$ is said to be a *linear combination* of the points $\boldsymbol{x}_i$, $i = 1, \ldots, n$, if there exist $\lambda_i \in \mathbb{R}$, $i = 1, \ldots, n$, such that $\boldsymbol{x} = \sum_{i=1}^n \lambda_i \boldsymbol{x}_i$. Depending on additional restrictions on the coefficients λ_i, other forms of combinations arise. If $\sum_{i=1}^n \lambda_i = 1$ then $\boldsymbol{x}$ is an *affine combination* of the points $\boldsymbol{x}_i$, $i = 1, \ldots, n$, and if additionally $\lambda_i \geq 0$, $i = 1, \ldots, n$, it becomes a *convex combination.* Hence, a convex combination is a particular affine combination and this is a particular linear combination. Since, by definition, a *linear subspace* is a set that includes all its linear combinations and an *affine manifold* is a set containing all its affine combinations, it follows that linear subspaces and affine manifolds are convex sets. We give the following compact definition.

Definition 2.2.4. *A set $\mathcal{A}$ is called* $\left\langle \begin{array}{c} \textit{a linear subspace} \\ \textit{an affine manifold} \\ \textit{a convex set} \end{array} \right\rangle$ *if* $\sum_{i=1}^n \lambda_i \mathcal{A} \subseteq \mathcal{A}$ *holds*

for each $n \in \mathbb{N}$ *with* $\left\langle \begin{array}{c} \lambda_i \in \mathbb{R} \\ \lambda_i \in \mathbb{R} \textit{ and } \sum_{i=1}^n \lambda_i = 1 \\ \lambda_i \geq 0 \textit{ and } \sum_{i=1}^n \lambda_i = 1 \end{array} \right\rangle$.

Examples of convex combinations of two points $\boldsymbol{x}_1$ and $\boldsymbol{x}_2$ are the points in the line segment $[\boldsymbol{x}_1, \boldsymbol{x}_2]$. An example of an affine combination given by the same points is the unique straight line that includes them, see Figure 2.4.

Figure 2.4: All convex combinations and "all" affine combinations of two points

Hulls are constructed externally by means of intersections, and internally by means of combinations. Thus, given a set $\mathcal{S}$ its *linear hull*, $\text{lin}(\mathcal{S})$, is the intersection of all the linear subspaces containing $\mathcal{S}$ (or the set of all linear combinations of elements in $\mathcal{S}$). Moreover, its *affine hull*, $\text{aff}(\mathcal{S})$ is the intersection of all affine manifolds containing $\mathcal{S}$ (or the set of all affine combinations of elements in $\mathcal{S}$). Finally, its *convex hull*, $\text{conv}(\mathcal{S})$, is the intersection of all convex sets containing $\mathcal{S}$ (or the set of all convex combinations of elements in $\mathcal{S}$).

The line in Figure 2.4 is a linear subspace if it includes $\mathbf{0}$, otherwise it is parallel to a unique linear subspace obtained, for instance, by $\text{aff}(\{\mathbf{0}, \boldsymbol{x}_2 - \boldsymbol{x}_1\})$. This is a consequence of a more general result mentioned in Section III.1.3 of Ref. [52] which states that

$$\text{aff}(\mathcal{S}) = \boldsymbol{x}_0 + \text{lin}(\mathcal{S} - \boldsymbol{x}_0) \tag{2.1}$$

for each $\boldsymbol{x}_0 \in \mathcal{S}$. It can also be shown that the corresponding linear subspace is unique and so the *dimension* of $\text{aff}(\mathcal{S})$ is defined as the dimension of $\text{lin}(\mathcal{S} - \boldsymbol{x}_0)$. Notice that $\mathcal{S} - \boldsymbol{x}_0$ is different from $\mathcal{S} \setminus \{\boldsymbol{x}_0\}$. The first is a *difference* while the second is an *exclusion*.

Figure 2.4 provides an example of a convex hull and of an affine hull. Also, the polytope in Figure 2.2 is the convex hull of its *vertices*, the "corner points" where two hyperplanes meet. A formal definition of vertex, or *extreme point*, corresponds to every point in a convex set that can not be written as a convex combination of any two other points in this set. Another example of a convex hull is provided by Figure 2.5 where the convex hull of the nonconvex set in the second picture in Figure 2.1 is represented as the set with a thick boundary.

When taking convex combinations in a set $\mathcal{S} \subseteq \mathbf{R}^s$ one only needs finitely many elements from $\mathcal{S}$ to represent any element of $\text{conv}(\mathcal{S})$. This result is clear for linear combinations since no more than s elements of a linear subspace of dimension s, a so-called *basis*, are necessary to generate all the others by linear combinations. Regarding affine manifolds, relation (2.1) implies that $x \in \text{aff}(\mathcal{S})$ equals $x_0 + \sum_{i=1}^{s} \lambda_i \epsilon_i$ with $\{\epsilon_1, \ldots, \epsilon_s\}$ some basis of the s-dimensional linear subspace $\text{lin}(\mathcal{S} - x_0)$.

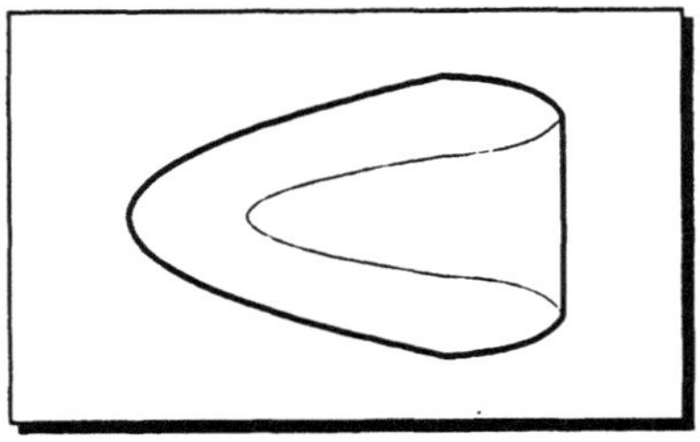

Figure 2.5: An example of a convex hull

Some rewriting leads to $x = (1 - \sum_{i=1}^{s} \lambda_i)x_0 + \sum_{i=1}^{s} \lambda_i(\epsilon_i + x_0)$ and this is an affine combination of $\{x_0, \epsilon_1 + x_0, \ldots, \epsilon_s + x_0\} \subseteq \text{aff}(\mathcal{S})$. Formally, the result for convex combinations, known as Carathéodory's theorem, is the following.

Theorem 2.2.1 (Theorem III.1.3.6 of Ref. [52]). *Any* $x \in \text{conv}(\mathcal{S}) \subseteq \mathbf{R}^s$ *can be represented as a convex combination of* $s + 1$ *elements of* $\mathcal{S}$.

The main difference between, on one hand, the linear and affine case and, on the other hand, the convex case is that the $s+1$ elements of $\mathcal{S}$ mentioned in Theorem 2.2.1 can not be fixed in advance. Hence, the terms *linear basis* and *affine basis* can be used but a "convex basis" does not exist in general.

An important consequence of Carathéodory's theorem is the following.

Theorem 2.2.2 (Theorem III.1.4.3 of Ref. [52]). *If* $\mathcal{S}$ *is bounded (respectively compact) then* $\text{conv}(\mathcal{S})$ *is bounded (respectively compact).*

Until now we only have considered halfspaces as special convex sets. However, a very important concept in convex analysis is the concept of a *cone*. A set $\mathcal{K}$ is called a cone if $\lambda\mathcal{K} \subseteq \mathcal{K}$ for every $\lambda > 0$. Observe that a set can be a cone without including $\mathbf{0}$. Moreover, since linear subspaces are cones, it is clear that not every cone has extreme points. However, if a cone has an extreme point it must be at

0. This unique extreme point of a cone is called the *apex*. Actually, as Section 2.3 illustrates, cones play in convexity the role of halfspaces in differential calculus.

Another important observation is that a cone does not need to be convex. For example, any pair of lines emanating from **0** form a cone without including the points in between them.

One could be tempted to define the *conical hull* of $\mathcal{S}$ as the intersection of all cones containing $\mathcal{S}$, i.e. the smallest cone containing $\mathcal{S}$. We are mostly interested in conical hulls of sets not including **0** in its interior, since it is easy to verify that any cone containing $\mathcal{S}$ has to equal $\mathbf{R}^s$ if $\mathbf{0} \in \text{int}(\mathcal{S})$. As examples of cones containing sets consider the sets represented in Figure 2.1 and put **0** somewhere outside these sets as in the following figure.

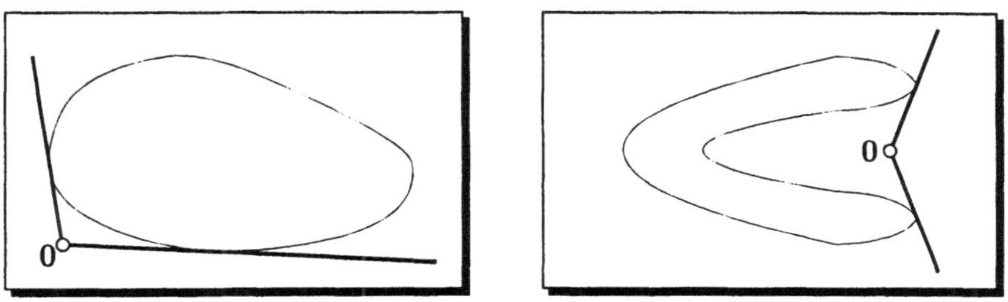

Figure 2.6: Examples of cones containing sets

Observe that the second picture in Figure 2.6 shows that the smallest cone containing a set does not have to be convex. However, if C is convex then the smallest cone containing C is also convex and it can be described as follows.

Theorem 2.2.3 (Corollary 2.6.3 of Ref. [89]). *If C is convex it follows that the set $\{\lambda \boldsymbol{x} : \lambda > 0,\ \boldsymbol{x} \in C\}$ is the smallest cone containing C.*

The essential cones in convex analysis are the convex cones and so the conical hull cone($\mathcal{S}$) of $\mathcal{S}$ is defined as the intersection of all convex cones containing $\mathcal{S}$ with **0** added to it. In order to give a construction "from the inside" define (Definition III.1.4.5 of Ref. [52]) a *conical combination* of elements $\{\boldsymbol{x}_1, \boldsymbol{x}_2, \ldots, \boldsymbol{x}_n\}$ as an element of the form $\sum_{i=1}^{n} \lambda_i \boldsymbol{x}_i$ with $\lambda_i \geq 0$. This is clearly a member of the previously presented family of combinations. Observe that it would be more adequate to call it a "convex conical combination" since any conical combination results from a convex combination by multiplying it by a nonnegative number. Hence, the conical hull cone($\mathcal{S}$) of $\mathcal{S}$ equals the set of all conical combinations of elements of $\mathcal{S}$. Observe first that cone($\mathcal{S}$) includes $\mathbf{0}$. On the other hand, we are mostly interested in conical hulls of convex sets and hence by Theorem 2.2.3 it follows that

$$\text{cone}(\mathcal{C}) := \{\lambda \boldsymbol{x} : \lambda > 0,\ \boldsymbol{x} \in \mathcal{C}\} \cup \mathbf{0}.$$

In general $\text{cl}(\text{cone}(\mathcal{C})) \neq \text{cone}(\text{cl}(\mathcal{C}))$. However, the following important result holds.

Lemma 2.2.1 (Proposition III.1.4.7 of Ref. [52]). *Let $\mathcal{S}$ be a nonempty compact set such that* $\mathbf{0} \notin \text{conv}(\mathcal{S})$. *Then* $\text{cl}(\text{cone}(\mathcal{S})) = \text{cone}(\mathcal{S}) = \text{cone}(\text{cl}(\mathcal{S}))$.

Associated with every nonempty convex cone $\mathcal{K}$ there is another important convex cone.

Definition 2.2.5 (Definition III.3.2.1 of Ref. [52]). *The polar cone of a nonempty convex cone $\mathcal{K}$ is given by*

$$\mathcal{K}^\circ := \{\boldsymbol{d} \in \mathbf{R}^s : \boldsymbol{d}^\top \boldsymbol{x} \leq 0 \textit{ for all } \boldsymbol{x} \in \mathcal{K}\}.$$

Polarity in this context can be understood as a generalization of normality. For instance, the polar cone of the halfspace $\mathcal{H}^{\leq}(\boldsymbol{a}, 0)$ is nothing else than the half line emanating from $\mathbf{0}$ and directed by $\boldsymbol{a}$. Reciprocally, if that half line is taken as the initial cone, its polar cone is $\mathcal{H}^{\leq}(\boldsymbol{a}, 0)$. Hence, we have just realized that if $\mathcal{K}$ is closed then the so-called *bipolar* given by $\mathcal{K}^{\circ\circ} = (\mathcal{K}^\circ)^\circ$ equals $\mathcal{K}$. Another observation is that $\mathcal{K}^\circ$ is always closed, even if $\mathcal{K}$ is not, and so we arrive at the following result.

Lemma 2.2.2 (Proposition III.4.2.7 of Ref. [52]). *If $\mathcal{K}$ is a nonempty convex cone then* $\mathcal{K}^{\circ\circ} = \text{cl}(\mathcal{K})$.

Another interesting characteristic of polarity is that it reverses inclusion.

$$\mathcal{K}_1 \subseteq \mathcal{K}_2 \ \Rightarrow\ \mathcal{K}_1^\circ \supseteq \mathcal{K}_2^\circ. \tag{2.2}$$

We can now introduce a cone which plays a prominent role in the sequel.

Definition 2.2.6 (Definition III.5.2.3 of Ref. [52]). *The direction $s \in \mathbf{R}^s$ is said to be normal to C at $x \in C$ if $s^{\mathsf{T}}(y-x) \leq 0$ for all $y \in C$. The set of all such directions is denoted by $\mathcal{N}_C(x)$ and it is called the normal cone to C at x.*

The following alternative characterization of a normal cone plays an important role in this book.

Lemma 2.2.3 (Proposition III.5.2.4 of Ref. [52]). $\mathcal{N}_C(x) = (\text{cone}(C - x))^\circ$.

Based on Lemma 2.2.3 we define the *normal cone* to C at x without requiring x to belong to C.

2.2.2. Topological properties of convex sets

It happens often that a convex set has no interior. As an example, consider a line segment in $\mathbf{R}^s$ for $s \geq 2$. However, by introducing a *relative* topology it turns out that convex sets always have a so-called *relative interior.* To be more precise, consider the affine hull $\text{aff}(C)$ of a convex set C. This affine manifold is parallel to a unique subspace and the *dimension* of the convex set C, $\dim(C)$, is defined as the dimension of this subspace. Hence, a convex set $C \subset \mathbf{R}^s$ is said to have *full dimension* if $\dim(C) = s$. Only for this case it follows that $\text{int}(C) \neq \emptyset$. For the general case we introduce the following definition.

Definition 2.2.7 (Definition III.2.1.1 of Ref. [52]).

$$\text{ri}(C) := \{x \in \text{aff}(C) : \text{aff}(C) \cap (x + \delta\mathcal{B}) \subseteq C \textit{ for some } \delta > 0\}.$$

Naturally, C is called *relatively open* if $\text{ri}(C)$ equals C. Observe that for full dimensional convex sets the relative interior equals the interior. Unlike the interior, the closure of C does not change if taken in the topology of $\mathbf{R}^s$ or in the relative topology of $\text{aff}(C)$. Hence, it makes no sense to define a relative closure. However, it makes sense to define a *relative boundary.*

Definition 2.2.8 (Remark III.2.1.2 of Ref. [52]). *For $C \subseteq \mathbf{R}^s$ a nonempty convex set its relative boundary* $\text{rbd}(C)$ *equals* $\text{cl}(C) \setminus \text{ri}(C)$.

The following results show the relevance of addressing the topological properties of convex sets relative to their affine hulls.

Theorem 2.2.4 (Theorem III.2.1.3 of Ref. [52]). *If C is nonempty and convex then* $\mathrm{ri}(C) \neq \emptyset$. *Moreover,* $\dim(\mathrm{ri}(C)) = \dim(C)$.

Lemma 2.2.4 (Lemma III.2.1.6 of Ref. [52]). *Let $x \in \mathrm{cl}(C)$ and $z \in \mathrm{ri}(C)$. Then* $]x, z] \subseteq \mathrm{ri}(C)$.

The previous lemma together with Theorem 6.4 of Ref. [89] yields the following result.

Theorem 2.2.5. *If C is a nonempty convex set then $z \in \mathrm{ri}(C)$ if and only if for every $x \in C$ there exists a $\mu < 0$ satisfying $z + \mu(x - z) \in \mathrm{ri}(C)$.*

Moreover, the relative interior of a convex set C is a very close representation of C itself. This is shown by the following pair of results.

Theorem 2.2.6 (Theorem 6.3 of Ref. [89]). *Every nonempty convex set C satisfies* $\mathrm{cl}(\mathrm{ri}(C)) = \mathrm{cl}(C)$ *and* $\mathrm{ri}(\mathrm{cl}(C)) = \mathrm{ri}(C)$.

Theorem 2.2.7 (Corollary 6.3.1 of Ref. [89]). *Let C_1 and C_2 be nonempty convex sets in $\mathbf{R}^s$. Then $\mathrm{cl}(C_1) = \mathrm{cl}(C_2)$ if and only if $\mathrm{ri}(C_1) = \mathrm{ri}(C_2)$. These conditions are equivalent to the condition that $\mathrm{ri}(C_1) \subseteq C_2 \subseteq \mathrm{cl}(C_1)$.*

Finally we give a characterization for the relative interior of a conical hull.

Theorem 2.2.8 (Corollary 6.8.1 of Ref. [89]). *If C is a nonempty convex set then* $\mathrm{ri}(\mathrm{cone}(C)) = \bigcup_{\lambda > 0} \lambda\, \mathrm{ri}(C)$.

2.2.3. Separation of convex sets

In this subsection we introduce the most important concept from convex analysis. Let C_1 and C_2 be two nonempty convex subsets of $\mathbf{R}^s$. These sets are said to be *properly separated* by $s \in \mathbf{R}^s$ if

$$\sup_{x_1 \in C_1} s^\top x_1 \le \inf_{x_2 \in C_2} s^\top x_2 \text{ and } \inf_{x_1 \in C_1} s^\top x_1 < \sup_{x_2 \in C_2} s^\top x_2.$$

We proceed directly to the most general *separation result.*

Theorem 2.2.9 (Theorem 11.3 of Ref. [89]). *The nonempty convex sets C_1 and C_2 can be properly separated by some $s \in \mathbf{R}^s$ if and only if* $\text{ri}(C_1) \cap \text{ri}(C_2) = \emptyset$.

We have just presented a very important result. Given two convex sets whose relative interiors do not intersect a third simple convex set, a hyperplane, can be "squeezed" between their relative interiors.

In the remainder, we are mostly interested in the case that one of the two sets reduces to a point x. Since $\text{ri}(\{x\}) = \{x\}$, the following result is an immediate consequence of Theorem 2.2.9.

Corollary 2.2.1. *The nonempty convex set C and the point x can be properly separated if and only if x is not an element of* $\text{ri}(C)$.

By construction, the separators of C and x form the set $\mathcal{N}_C(x)$ which by Lemma 2.2.3 equals $(\text{cone}(C - x))^\circ$.

This basic observation is responsible for the developments in Chapter 5.

2.3. Convex functions

We recall first that a function $f : \mathbf{R}^s \longrightarrow \overline{\mathbf{R}}$ is called *proper* if the *domain* of f, given by $\text{dom}(f) := \{x \in \mathbf{R}^s : f(x) < \infty\}$, is nonempty and $f(x) > -\infty$ for every $x \in \mathbf{R}^s$. In order to relate functions to sets, we introduce the *epigraph* of a function $f : \mathbf{R}^s \longrightarrow \overline{\mathbf{R}}$. This is the subset of $\mathbf{R}^{s+1}$ given by

$$\text{epi}(f) := \{(x, \alpha) \in \mathbf{R}^{s+1} : \alpha \ge f(x)\},$$

i.e. the set of "everything above the graph" of f. We also mention that f is called *closed* if its epigraph is a closed set. A closed function is also called *lower semicontinuous.*

A function f is called *convex* if its epigraph is a convex set. This, together with the definition of epigraph, leads to f being convex if and only if $[(x, f(x)), (y, f(y))] \subseteq \text{epi}(f)$ for every $x, y \in \mathbf{R}^s$, or (for a proper f) to the perhaps more familiar definition (Ref. [52, 89])

$$f(\lambda x + (1-\lambda)y) \leq \lambda f(x) + (1-\lambda)f(y)$$

for every $x, y \in \mathbf{R}^s$ and every $0 < \lambda < 1$.

2.3.1. Continuity of convex functions

Convex functions have nice continuity properties. The next result makes this statement more precise.

Theorem 2.3.1 (Theorem 10.1 of Ref. [89]). *A proper convex function $f : \mathbf{R}^s \longrightarrow \overline{\mathbf{R}}$ is continuous on every relatively open convex subset of its domain. In particular, f is continuous on* ri(dom(f)).

On compact subsets of ri(dom(f)) a stronger property holds. This property is now defined.

Definition 2.3.1 (Ref. [76]). *A proper function $f : \mathbf{R}^s \longrightarrow \overline{\mathbf{R}}$ is called Lipschitz continuous on a set $S \subseteq$ dom(f) if a constant L_f exists satisfying $|f(z) - f(y)| \leq L_f \|z - y\|_2$ for every z, y belonging to S. Moreover, L_f is called the Lipschitz constant of f on S.*

Theorem 2.3.2 (Theorem 10.4 of Ref. [89]). *A proper convex function $f : \mathbf{R}^s \longrightarrow \overline{\mathbf{R}}$ is Lipschitz continuous on any compact subset of* ri(dom(f)).

The following result is now an immediate consequence of the previous pair of results.

Corollary 2.3.1. *A finite-valued convex function $f : \mathbf{R}^s \longrightarrow \mathbf{R}$ is continuous on $\mathbf{R}^s$ and Lipschitz continuous on every compact subset of $\mathbf{R}^s$.*

2.3.2. Lower level sets and the subdifferential

By definition, a convex function $f : \mathbf{R}^s \longrightarrow \overline{\mathbf{R}}$ is associated with its epigraph, which is a convex subset of $\mathbf{R}^{s+1}$. However, associated with each convex function, there exist other important convex sets. These convex sets are subsets of the domain of f and play an important role in optimization. Since we are mainly concerned with minimizing functions, the "important" sets are the so-called *lower level sets.* The lower level set of a function f of level $\alpha \in \mathbf{R}$ is defined by $\mathcal{L}_f^{\leq}(\alpha) := \{x \in \mathbf{R}^s : f(x) \leq \alpha\}$. Two important subsets of this set also deserve attention. The so-called *strict lower level set* of level $\alpha \in \mathbf{R}$ given by $\mathcal{L}_f^{<}(\alpha) := \{x \in \mathbf{R}^s : f(x) < \alpha\}$ and the so-called *level set* of level $\alpha \in \mathbf{R}$ given by $\mathcal{L}_f^{=}(\alpha) := \{x \in \mathbf{R}^s : f(x) = \alpha\}$. The first two sets, lower and strict lower level, are convex sets for every $\alpha \in \mathbf{R}$ if f is a convex function. Moreover, since proper convex functions are continuous in the relative interior of their domain (Theorem 2.3.1), these sets are respectively closed and open.

A less obvious, but very important, property of level sets of convex functions is the following.

Lemma 2.3.1 (Proposition VI.1.3.3 of Ref. [52]). *If $g : \mathbf{R}^s \longrightarrow \overline{\mathbf{R}}$ is a proper convex function satisfying $g(\widehat{x}) < 0$ for some $\widehat{x} \in \mathbf{R}^s$ then $\mathrm{cl}(\mathcal{L}_g^{<}(0)) = \mathcal{L}_g^{\leq}(0)$, $\mathrm{int}(\mathcal{L}_g^{\leq}(0)) = \mathcal{L}_g^{<}(0)$ and $\mathrm{bd}(\mathcal{L}_g^{\leq}(0)) = \mathcal{L}_g^{=}(0)$.*

Important instances of level sets are those associated with *affine* functions (often called "linear" as a consequence of the popularity of the term *linear programming*) as $h(x) := \boldsymbol{a}^{\top}x + b$ with $\boldsymbol{a} \in \mathbf{R}^s$ and $b \in \mathbf{R}$. Each level set of such a function is a hyperplane, $\mathcal{L}_h^{=}(\alpha) = \mathcal{H}^{=}(\boldsymbol{a}, \alpha - b)$, and each lower level set is a lower halfspace, $\mathcal{L}_h^{\leq}(\alpha) = \mathcal{H}^{\leq}(\boldsymbol{a}, \alpha - b)$. This offers another justification for the term lower halfspace. If an affine function satisfies $h(\mathbf{0}) = 0$ then it is called *linear.*

Perhaps the most important family of convex sets associated with a convex function is the so-called *subdifferential* or *subgradient set.* Before defining this set we give a definition of *subgradient.* A vector $x^{\star} \in \mathbf{R}^s$ is called a subgradient of f at x if the so-called *subgradient inequality*

$$f(y) \geq f(x) + x^{\star\top}(y - x)$$

holds for every $y \in \mathbf{R}^s$. The set of all subgradients at x is called the subdifferential of f at x and denoted by $\partial f(x)$. The next result discusses for which points the subdifferential is a nonempty set. Observe that it is easy to verify that the subdifferential is a closed and convex set.

Theorem 2.3.3 (Theorem 23.4 of Ref. [89]). *If $f : \mathbf{R}^s \longrightarrow \overline{\mathbf{R}}$ is a proper convex function then for every $x \in$ ri(dom(f)) the subdifferential $\partial f(x)$ is nonempty. In particular, if f is finite-valued then $\partial f(x)$ is nonempty for every $x \in \mathbf{R}^s$.*

Interesting properties of the subdifferential are presented after introducing in the next subsection a very important class of convex functions.

2.3.3. Sublinear functions and directional derivatives

We already mentioned some special convex functions: affine and linear functions. Now we define *sublinear functions.*

Definition 2.3.2 (Definition V.1.1.1 of Ref. [52]). *A function $\sigma : \mathbf{R}^s \longrightarrow \overline{\mathbf{R}}$ is called sublinear if it is proper, convex and positively homogeneous, i.e. $\sigma(tx) = t\sigma(x)$ for every x and every $t > 0$.*

These functions have interesting epigraphs: their epigraphs are cones in $\mathbf{R}^{s+1}$. As Proposition V.1.1.4 of Ref. [52] shows, an alternative characterization of sublinearity is $\sigma(\alpha x + \beta y) \leq \alpha\sigma(x) + \beta\sigma(y)$ for every x, y and for every $\alpha, \beta > 0$. Observe that this justifies the term "sublinear". A function $\sigma : \mathbf{R}^s \longrightarrow \overline{\mathbf{R}}$ is called *subadditive* if it is proper and $\sigma(x+y) \leq \sigma(x)+\sigma(y)$ holds for every x and y. The same proposition gives a second alternative characterization: a function is sublinear if and only if it is subadditive and positively homogeneous. Finally we observe (Ref. [52, 89]) that $\partial\sigma(x) \subseteq \partial\sigma(0)$ for every $x \in \mathbf{R}^s$.

Proper convex functions also have *directional derivatives* at every point belonging to their domain (Theorem 23.1 of Ref. [89]). Moreover, if the directional derivative is taken at some point in the relative interior of the domain then the directional derivative, regarded as a function of the direction, is a closed sublinear function.

Lemma 2.3.2 (Proposition VI.1.1.2 of Ref. [52] and Theorem 23.4 of Ref. [89]). *If $f : \mathbf{R}^s \longrightarrow \overline{\mathbf{R}}$ is a proper convex function then for each x belonging to* ri(dom(f)) *the directional derivative*

$$f'(x;d) := \lim_{t \downarrow 0} \frac{f(x+td) - f(x)}{t}$$

exists and $d \longmapsto f'(x;d)$ is a closed and sublinear function. Moreover, if f is a finite-valued convex function then the function $d \longmapsto f'(x;d)$ is finite-valued and sublinear (hence continuous) at every $x \in \mathbf{R}^s$.

Recall that f is *differentiable* at x if and only if $d \longmapsto f'(x;d)$ is linear. Moreover, by Corollary VI.2.1.4 of Ref. [52], a convex function f is differentiable at x if and only if $\partial f(x)$ is a *singleton*, i.e. a set composed by only one element. If that is the case then $\partial f(x) = \{\nabla f(x)\}$. A remarkable result is the following.

Theorem 2.3.4 (Theorem IV.4.2.3 of Ref. [52]). *If $f : \mathbf{R}^s \longrightarrow \overline{\mathbf{R}}$ is a proper convex function then the points of* int(dom(f)) *where it fails to be differentiable form a set of zero (Lebesgue) measure.*

Hence, if we restrict ourselves to finite-valued convex functions, these functions are continuous everywhere and differentiable "almost" everywhere.

2.3.4. Support functions and gauges

Given a nonempty set $\mathcal{S}$, the function $\sigma_{\mathcal{S}} : \mathbf{R}^s \longrightarrow \overline{\mathbf{R}}$ given by

$$\sigma_{\mathcal{S}}(x) := \sup\{s^{\mathsf{T}}x : s \in \mathcal{S}\}$$

is called the *support function* of $\mathcal{S}$. Since $\sigma_{\mathcal{S}}(\mathbf{0}) = 0$ and $\sigma_{\mathcal{S}}(x) > -\infty$ for every $x \in \mathbf{R}^s$ it follows that $\sigma_{\mathcal{S}}$ is proper. Regarding $\mathcal{S}$ as an index set it is immediately clear that $\sigma_{\mathcal{S}}$ is closed and convex. Moreover, by Proposition V.2.1.2 of Ref. [52] it turns out that it is sublinear, regardless of the nature of $\mathcal{S}$. This remarkable fact exhibits a fundamental duality correspondence between sublinear functions and convex sets.

Theorem 2.3.5 (Theorem V.3.1.1 of Ref. [52]). *If σ is a closed sublinear function then σ is the support function of the nonempty closed convex set $\mathcal{S}_\sigma := \{s \in \mathbb{R}^s : s^\top d \leq \sigma(d)$ for every $d \in \mathbb{R}^s\}$.*

Corollary 2.3.2 (Corollary 13.2.2 of Ref. [89]). *The support functions of the non-empty bounded convex sets are the finite sublinear functions.*

We already found in the previous subsection an important closed sublinear function, namely the directional derivative of a convex function at some point in the relative interior of its domain. The next result shows that the set it supports is nothing else than the subdifferential. This result can be easily obtained from Theorem 2.3.5, Corollary 2.3.2 and Lemma 2.3.2.

Theorem 2.3.6. *If $f : \mathbb{R}^s \longrightarrow \overline{\mathbb{R}}$ is a proper convex function then the directional derivative $d \longmapsto f'(x; d)$ with $x \in \mathrm{ri}(\mathrm{dom}(f))$ is the support function of the subdifferential $\partial f(x)$, i.e.*

$$f'(x; d) = \sup\{x^{\star\top} d : x^\star \in \partial f(x)\}.$$

In particular, if f is finite-valued then $\partial f(x)$ is a compact convex set and $d \longmapsto f'(x; d)$ a finite-valued sublinear function for every $x \in \mathbb{R}^s$.

For finite-valued convex functions a consequence of Lemma 2.3.1 and the previous result is of utmost importance: every normal direction to the lower level set is a nonnegative multiple of a subgradient (Theorem VI.1.3.5 of Ref. [52]). This is clearly stated as $\mathcal{N}_{\mathcal{L}_f^\leq}(x) = \mathrm{cone}(\partial f(x))$.

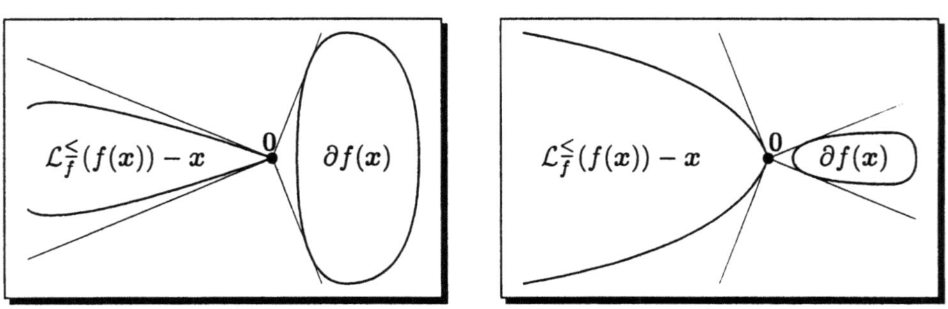

Figure 2.7: Relation between the lower level set and the subdifferential

An interpretation of this result is provided by Figure 2.7. Notice in this figure that not every subgradient is a negative multiple of a descent direction from the point x. Take, for instance, in the first picture any of the two subgradients corresponding to the tangent points to the boundary of the conical hull of $\partial f(x)$. Observe, however, that in the second picture this conical hull is a much "narrower" cone for which every subgradient is symmetric to a descent direction. If the subdifferential reduces to one point, the gradient of f at x, then this vector is well-known to yield by symmetry the *steepest descent direction* of f from x. Moreover, the conical hull of $\mathcal{L}_f^{\lneq}(f(x)) - x$ becomes an halfspace and this illustrates how cones play the role of halfspaces when convexity replaces differentiability, as mentioned in Section 2.2.1.

The following result, although an immediate consequence of the definition of subgradient and Theorem 2.3.6, is extremely important.

Theorem 2.3.7 (Theorem VI.2.2.1 of Ref. [52]). *If $f : \mathbf{R}^s \longrightarrow \mathbf{R}$ is convex then the following three properties are equivalent.*

1. *x is a global minimum of f, i.e. $f(y) \geq f(x)$ for every $y \in \mathbf{R}^s$.*

2. *$\mathbf{0} \in \partial f(x)$.*

3. *$f'(x; \mathbf{d}) \geq 0$ for every $\mathbf{d} \in \mathbf{R}^s$.*

We now define a *gauge*. Later we show that such a function is also a support function. Let $\mathcal{G} \subseteq \mathbf{R}^s$ be a compact convex set with $\mathbf{0} \in \text{int}(\mathcal{G})$. A gauge or *Minkowsky functional* (Ref. [52, 73, 89]) is defined for each $x \in \mathbf{R}^s$ by

$$\gamma_{\mathcal{G}}(x) := \inf\{t > 0 : x \in t\mathcal{G}\}. \tag{2.3}$$

If $\mathcal{G}$ is symmetric around $\mathbf{0}$ the corresponding gauge is called a *norm*.

By looking at (2.3) one sees immediately that $\mathcal{G} = \mathcal{L}_{\gamma_{\mathcal{G}}}^{\leq}(1)$. Although more general gauges may be defined by having $\mathbf{0}$ in $\mathcal{G}$ and not necessarily in $\text{int}(\mathcal{G})$ we are only interested in finite-valued gauges. By Theorem V.1.2.5 of Ref. [52] these gauges are those for which $\mathbf{0} \in \text{int}(\mathcal{G})$. Moreover, such a gauge is a nonnegative continuous sublinear function (Theorem V.1.2.5 of Ref. [52]) and hence its epigraph is a closed cone. This justifies the following strategy to "construct" the epigraph of a gauge. Consider a replica of $\mathcal{G}$ shifted "upwards" one unit, i.e. consider $\mathcal{G} \times \{1\} \subseteq \mathbb{R}^{s+1}$. Now, $\gamma_{\mathcal{G}}$ is just the function whose epigraph is the conical hull of this set in $\mathbb{R}^{s+1}$, i.e. $\text{epi}(\gamma_{\mathcal{G}}) = \text{cone}(\mathcal{G} \times \{1\})$. Figure 2.8 illustrates this construction for the most well-known gauge, the Euclidean norm, i.e. for $\mathcal{G} = \{x \in \mathbb{R}^s : \|x\|_2 \leq 1\}$.

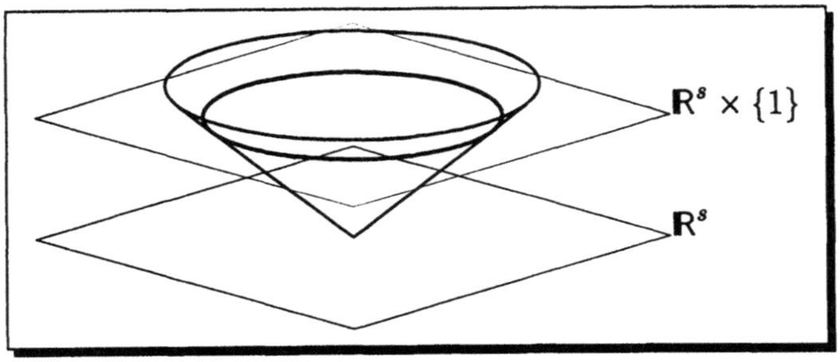

Figure 2.8: Construction of a gauge

In order to present an extremely important property of gauges we need the following definition.

Definition 2.3.3 (Ref. [89]). *Let $\mathcal{G}$ be a convex set. The polar set of $\mathcal{G}$ is given by*

$$\mathcal{G}^{\circ} := \{d \in \mathbb{R}^s : d^{\mathsf{T}} x \leq 1 \text{ for all } x \in \mathcal{G}\}.$$

Since $\gamma_{\mathcal{G}}$ is a continuous sublinear function one may derive the first part of the following result from Theorem 2.3.5 or Corollary 2.3.2. The other part is discussed in Example VI.3.2 of Ref. [52].

Theorem 2.3.8 (Ref. [52, 89]). *If $\mathcal{G}$ is a nonempty convex set with $\mathbf{0} \in \text{int}(\mathcal{G})$ then*

$$\gamma_{\mathcal{G}}(x) = \sigma_{\mathcal{G}^{\circ}}(x) := \sup\{x^{\star\mathsf{T}} x : x^{\star} \in \mathcal{G}^{\circ}\}. \tag{2.4}$$

Moreover, for every $x_0 \in \mathbb{R}^s$ the subdifferential set $\partial\gamma_{\mathcal{G}}(x_0)$ is given by

$$\partial\gamma_{\mathcal{G}}(x_0) = \{x_0^{\star} \in \mathcal{G}^{\circ} : x_0^{\star\mathsf{T}} x_0 = \gamma_{\mathcal{G}}(x_0)\}.$$

In the next pair of examples we discuss some well-known gauges.

Example 2.3.1 (ℓ_p-norm). *Let $n_p : \mathbf{R}^s \longrightarrow \mathbf{R}_+$ be given by $n_p(\boldsymbol{x}) := \|\boldsymbol{x}\|_p$ with $\|\boldsymbol{x}\|_p$ the ℓ_p-norm, $1 < p < \infty$, and suppose $\mathcal{G}_p := \{\boldsymbol{x} \in \mathbf{R}^s : n_p(\boldsymbol{x}) \leq 1\}$. By the Hölder inequality (Ref. [106]) it follows that $\mathcal{G}_p^\circ = \mathcal{G}_q := \{\boldsymbol{x} \in \mathbf{R}^s : n_q(\boldsymbol{x}) \leq 1\}$ with $\frac{1}{p} + \frac{1}{q} = 1$. Moreover, it can be easily verified that the function n_p is differentiable outside $\{\mathbf{0}\}$ and so its subdifferential $\partial n_p(\boldsymbol{x})$, $\boldsymbol{x} \neq \mathbf{0}$, only consists of its gradient $\nabla n_p(\boldsymbol{x})$ (Ref. [89]). Observe that this gradient $\nabla n_p(\boldsymbol{x})$, $\boldsymbol{x} \neq \mathbf{0}$, is given by*

$$\nabla n_p(\boldsymbol{x}) = \|\boldsymbol{x}\|_p^{1-p} \begin{bmatrix} \operatorname{sign}(x_1)|x_1|^{p-1} \\ \vdots \\ \operatorname{sign}(x_n)|x_n|^{p-1} \end{bmatrix}$$

with sign *the sign function defined as*

$$\operatorname{sign}(x) := \begin{cases} -1 & \text{if } x < 0 \\ 0 & \text{if } x = 0 \\ 1 & \text{if } x > 0 \end{cases}$$

while $\partial n_p(\mathbf{0}) = \mathcal{G}_p^\circ = \mathcal{G}_q$.

Example 2.3.2 (Polyhedral gauges). *Let $\mathcal{G} \subseteq \mathbf{R}^s$ be a polytope having $\mathbf{0}$ in its interior. In this case $\mathcal{G} = \operatorname{conv}(\{\boldsymbol{v}_1, \ldots, \boldsymbol{v}_r\})$ with $\boldsymbol{v}_1, \ldots, \boldsymbol{v}_r$ its vertices. Hence, $\mathcal{G}^\circ$ equals $\{\boldsymbol{x}^\star \in \mathbf{R}^s : \boldsymbol{x}^{\star\top}\boldsymbol{v}_i \leq 1$ for every $i = 1, \ldots, r\}$ and so $\mathcal{G}^\circ$ is again a polytope with vertices $\boldsymbol{v}_1^\circ, \ldots, \boldsymbol{v}_{r^\circ}^\circ$. Observe that the $(s-1)$-dimensional facets $\{\boldsymbol{x}^\star \in \mathbf{R}^s : \boldsymbol{x}^{\star\top}\boldsymbol{v}_i = 1\}$ of $\mathcal{G}^\circ$ correspond uniquely to the vertices of $\mathcal{G}$. In particular, for $s = 2$ we have $r^\circ = r$ and the vertices of $\mathcal{G}^\circ$ can be obtained as follows. First number the vertices of $\mathcal{G}$ clockwise and then compute the intersection of the 1-dimensional facets $\boldsymbol{x}^{\star\top}\boldsymbol{v}_i = 1$ and $\boldsymbol{x}^{\star\top}\boldsymbol{v}_{i+1} = 1$ of $\mathcal{G}^\circ$ for $i = 1, \ldots, r-1$ and the intersection of $\boldsymbol{x}^{\star\top}\boldsymbol{v}_i = 1$ with $\boldsymbol{x}^{\star\top}\boldsymbol{v}_1 = 1$ for $i = r$. These intersection points are the vertices of $\mathcal{G}^\circ$. By Theorem 2.3.8 we obtain now for $s \geq 2$ that*

$$\gamma_{\mathcal{G}}(\boldsymbol{x}) = \sup\{\boldsymbol{x}^{\star\top}\boldsymbol{x} : \boldsymbol{x}^\star \in \mathcal{G}^\circ\} = \max\{\boldsymbol{v}_i^{\circ\top}\boldsymbol{x} : i = 1, \ldots, r^\circ\}.$$

Introducing $I(\boldsymbol{x}_0) := \{1 \leq i \leq r^\circ : \gamma_{\mathcal{G}}(\boldsymbol{x}_0) = \boldsymbol{v}_i^{\circ\top}\boldsymbol{x}_0\}$ it follows again by Theorem 2.3.8 that

$$\partial\gamma_{\mathcal{G}}(\boldsymbol{x}_0) = \operatorname{conv}\{\boldsymbol{v}_i^\circ : i \in I(\boldsymbol{x}_0)\}.$$

Finally, observe by Theorem 1 of Ref. [109] that any polyhedral gauge $\gamma_{\mathcal{G}}$ with $\mathcal{G}$ symmetric yields a norm that computes the distance between two points as the fastest path in a dense highway grid with roads in every direction v_i, $i = 1, \ldots, r$ and speed equal to $\|v_i\|_2$ in each road. This result also holds in the asymmetric case and for an illustration we refer to Section 4.2.

Observe that Theorem 2.3.8 gives a fundamental relation between support functions and gauges. It also enables the support function of a compact convex set $\mathcal{C}$ including $\mathbf{0}$ in its interior to be geometrically constructed in a similar fashion as a gauge is. This construction proceeds as follows. Take $\mathcal{C}$ and construct its gauge as in Figure 2.8 by taking the function whose epigraph equals cone($\mathcal{C} \times \{1\}$). Now take the symmetric of this set (i.e. $-$ cone($\mathcal{C} \times \{1\}$)) and the required support function is just the function having as its epigraph the polar cone of this set. See Figure 2.9 for an illustration.

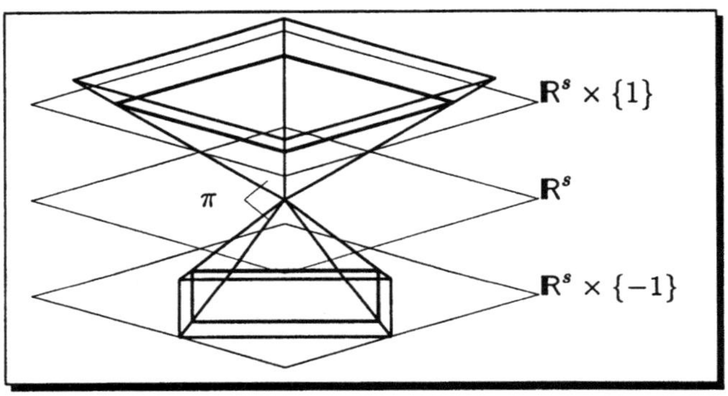

Figure 2.9: Relation between a support function and a gauge

Notice, since $\mathcal{C}^\circ = \mathcal{L}^{\leq}_{\sigma_{\mathcal{C}}}(1)$, that this also illustrates the construction of $\mathcal{C}^\circ$ when $\mathbf{0} \in \text{int}(\mathcal{C})$.

2.3.5. Calculus rules with subdifferentials

It is well-known from mathematical analysis (Ref. [92]) that the definition of gradient is not very useful to actually compute gradients. Instead, so-called ***calculus rules*** exist which allow the derivation of gradients of "complicated" functions by means of gradients of "simple" functions used as building blocks. Most of these rules have counterparts when gradients are replaced by subgradients. We now recall the most important ones.

Theorem 2.3.9 (Theorem VI.4.1.1 of Ref. [52]). *If $f, g : \mathbf{R}^s \longrightarrow \mathbf{R}$ are convex functions and α and β are nonnegative scalars then $\partial(\alpha f + \beta g)(x) = \alpha\partial f(x) + \beta\partial g(x)$ for every x.*

It is well known that the above property holds for differential functions regardless of the sign of α and β. For subgradients it does not hold for negative scalars since the difference of convex functions is not convex. For this reason, the set of convex functions is a cone while the set of differentiable functions is a linear space. An important generalization of the previous result, which amounts to a post-composition of a vector of two convex functions with a nondecreasing linear function, is the following.

Theorem 2.3.10 (Theorem VI.4.3.1 of Ref. [52]). *Let $f_i : \mathbf{R}^s \longrightarrow \mathbf{R}$ be convex functions, $1 \le i \le n$, and $\boldsymbol{f} : \mathbf{R}^s \longrightarrow \mathbf{R}^n$ the vector function whose components are the f_i, i.e. $\boldsymbol{f}(x) := (f_1(x), \ldots, f_n(x))$. If $g : \mathbf{R}^n \longrightarrow \mathbf{R}$ is convex and componentwise nondecreasing, i.e. $x \le y \Rightarrow g(x) \le g(y)$, then*

$$\partial(g \circ \boldsymbol{f})(x) = \operatorname{conv}\left(\sum_{i=1}^{n} \xi_i^\star x_i^\star : \boldsymbol{\xi}^\star \in \partial g(\boldsymbol{f}(x)),\ x_i^\star \in \partial f_i(x),\ 1 \le i \le n\right).$$

Notice that if $g : \mathbf{R}^n \longrightarrow \mathbf{R}$ is differentiable then Theorem 2.3.10 simplifies to

$$\partial(g \circ \boldsymbol{f})(x) = \sum_{i=1}^{n} \tfrac{\partial g}{\partial z_i}(\boldsymbol{f}(x))\partial f_i(x).$$

Another extremely important corollary (known as the Dubovitskiĭ-Milyutin theorem) of this general composition result shows that the cone of convex functions is closed under the finite max operator, while the space of differentiable functions is not.

Corollary 2.3.3 (Corollary VI.4.3.2 of Ref. [52]). *Let $f_i : \mathbf{R}^s \longrightarrow \mathbf{R}$ be convex functions, $1 \le i \le n$, and define $f(x) := \max_{1 \le i \le n} f_i(x)$. Denoting the so-called active* *set by* $I(x) := \{i : f_i(x) = f(x),\ 1 \leq i \leq n\}$ *we have*

$$\partial f(x) = \operatorname{conv}\left(\bigcup_{i \in I(x)} \partial f_i(x)\right).$$

This concludes our small introduction to convex functions. The next section discusses more general functions.

2.4. Quasiconvex functions

Quasiconvex functions are those functions whose lower level sets are convex sets. An example is provided by Figure 2.10 showing a function having only concentric circles as lower level sets. Observe, since its epigraph is not convex, that this is not a convex function. However, this is a "nice" quasiconvex function, satisfying Lemma 2.3.1 and Theorem 2.3.7. Most of the quasiconvex functions do not satisfy this result and are responsible for a significant part of Chapter 5.

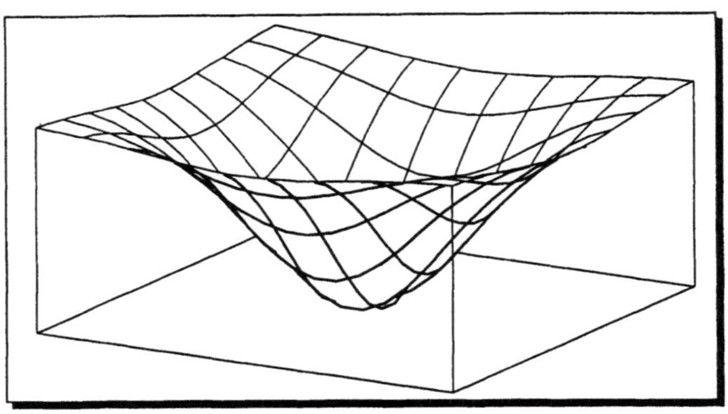

Figure 2.10: A bivariate quasiconvex function

A finite-valued quasiconvex function does not need to be continuous. For instance, the sign function defined in Example 2.3.1 is quasiconvex and neither upper nor lower semicontinuous in 0. The same function also exemplifies a quasiconvex function neither satisfying Lemma 2.3.1 nor Theorem 2.3.7.

Clearly, for f quasiconvex, and hence $\mathcal{L}_f^{\leq}(\alpha)$ convex for every $\alpha \in \mathbf{R}$, it follows that $\mathrm{dom}(f) = \bigcup_{\alpha \in \mathbf{R}} \mathcal{L}_f^{\leq}(\alpha)$ is convex due to $\mathcal{L}_f^{\leq}(\alpha) \subseteq \mathcal{L}_f^{\leq}(\beta)$ for every $\alpha \leq \beta$.

In the following result some well-known equivalent characterizations of quasiconvexity are listed.

Lemma 2.4.1 (Ref. [85]). *The following conditions are equivalent.*

1. *The function* $f : \mathbf{R}^s \longrightarrow \overline{\mathbf{R}}$ *is quasiconvex.*
2. *The strict lower level sets* $\mathcal{L}_f^{<}(\alpha) := \{x \in \mathbf{R}^s : f(x) < \alpha\}$ *are convex for every* $\alpha \in \mathbf{R}$.
3. *For every* $x, y \in \mathbf{R}^s$ *and* $0 < \lambda < 1$ *it follows that*

$$f(\lambda x + (1-\lambda)y) \leq \max\{f(x), f(y)\}.$$

Definition 2.4.1 (Ref. [7]). *Let* $C \subseteq \mathbf{R}^s$ *be a convex set. A function* $f : \mathbf{R}^s \longrightarrow \overline{\mathbf{R}}$ *is called quasiconvex on* C *if the sets* $\mathcal{L}_f^{\leq}(\alpha) \cap C$ *are convex for every* $\alpha \in \mathbf{R}$.

Observe (Ref. [7]) that $f : \mathbf{R}^s \longrightarrow \overline{\mathbf{R}}$ is quasiconvex on C if and only if for every $x_1, x_2 \in C$ and $0 < \lambda < 1$ the inequality $f(\lambda x_1 + (1-\lambda)x_2) \leq \max\{f(x_1), f(x_2)\}$ holds. Important quasiconvex functions are now defined.

Definition 2.4.2 (Ref. [77]). *A function* $f : \mathbf{R}^s \longrightarrow \overline{\mathbf{R}}$ *is called evenly quasiconvex if its lower level sets are evenly convex.*

Another class of functions closely related to quasiconvexity and Lipschitz continuity is now given by the next definition.

Definition 2.4.3 (Ref. [80]). *A function* $f : \mathbf{R}^s \longrightarrow \mathbf{R}$ *is called lower subdifferentiable on* $\mathcal{S} \subseteq \mathbf{R}^s$ *if for every* $z \in \mathcal{S}$ *there exists some* $z^\star \in \mathbf{R}^s$ *satisfying*

$$f(y) \geq f(z) + z^{\star\mathsf{T}}(y - z)$$

for every y *belonging to* $\mathcal{L}_f^{<}(f(z)) \cap \mathcal{S}$. *The set of all so-called lower subgradients* $z^\star$ *of the function* f *at* z *is denoted by* $\partial^- f(z)$. *Finally, the function* $f : \mathbf{R}^s \longrightarrow \mathbf{R}$

is called boundedly lower subdifferentiable on $\mathcal{S}$ if f is lower subdifferentiable on $\mathcal{S}$ and there exists a constant bound $N > 0$ such that for every $z \in \mathcal{S}$ there exists some $z^\star \in \partial^- f(z)$ with $\|z^\star\|_2 \leq N$.

It follows from Theorem 2.3 of Ref. [80] that if $f : \mathbb{R}^s \longrightarrow \mathbb{R}$ is quasiconvex and Lipschitz continuous on $\mathbb{R}^s$ with Lipschitz constant L_f then it is boundedly lower subdifferentiable on $\mathbb{R}^s$ with bound L_f.

2.5. Other directional derivatives

In this last section of the present chapter we consider general results on directional derivatives and some of their generalizations.

Lemma 2.5.1 (Ref. [19]). *If $f : \mathbb{R}^s \longrightarrow \mathbb{R}$ is given by $f(\boldsymbol{x}) := \max\{x_1, \ldots, x_n\}$ then the directional derivative $f'(\boldsymbol{x}; \boldsymbol{d})$ exists and is finite-valued for every $\boldsymbol{x}, \boldsymbol{d} \in \mathbb{R}^s$. Moreover, it follows that*

$$f'(\boldsymbol{x}; \boldsymbol{d}) = \max\{d_i : i \in I(\boldsymbol{x})\}$$

with $I(\boldsymbol{x}) := \{1 \leq i \leq n : f(\boldsymbol{x}) = x_i\}$.

If a directional derivative does not exist then one may use the following generalization.

Definition 2.5.1 (Ref. [60]). *If $f : \mathbb{R}^s \longrightarrow \overline{\mathbb{R}}$ is some function with $f(\boldsymbol{x})$ finite the Dini upper derivative of f at $\boldsymbol{x}_0$ in the direction $\boldsymbol{d}$ is given by*

$$f'_+(\boldsymbol{x}; \boldsymbol{d}) := \limsup_{t \downarrow 0} \frac{f(\boldsymbol{x} + t\boldsymbol{d}) - f(\boldsymbol{x})}{t}.$$

In order to prove the next composition result we need to introduce the following generalization of Lipschitz continuity.

Definition 2.5.2 (Ref. [19]). *A function $f : \mathbb{R}^s \longrightarrow \overline{\mathbb{R}}$ is called locally Lipschitz if for every $\boldsymbol{x} \in \operatorname{int}(\operatorname{dom}(f))$ there exists some $\varepsilon > 0$ and a Lipschitz constant $0 < L_{\boldsymbol{x}} < \infty$ such that $|f(\boldsymbol{z}) - f(\boldsymbol{y})| \leq L_{\boldsymbol{x}} \|\boldsymbol{z} - \boldsymbol{y}\|_2$ for every $\boldsymbol{z}, \boldsymbol{y} \in \boldsymbol{x} + \varepsilon\mathcal{B}$.*

Lemma 2.5.2. *If $f : \mathbb{R}^s \longrightarrow \mathbb{R}$ is locally Lipschitz and the component functions $u_i : \mathbb{R}^s \longrightarrow \mathbb{R}$, $i = 1, \ldots, n$, of $\boldsymbol{u}(\boldsymbol{x}) := (u_1(\boldsymbol{x}), \ldots, u_n(\boldsymbol{x}))$ have finite Dini upper derivatives then the composite function $g : \mathbb{R}^s \longrightarrow \mathbb{R}$ given by $g(\boldsymbol{x}) := f(\boldsymbol{u}(\boldsymbol{x}))$ has the following Dini upper derivative*

$$g'_+(\boldsymbol{x}; \boldsymbol{d}) = f'_+\left(\boldsymbol{u}(\boldsymbol{x}); \boldsymbol{u}'_+(\boldsymbol{x}; \boldsymbol{d})\right)$$

with $\boldsymbol{u}'_+(\boldsymbol{x}; \boldsymbol{d}) := (u'_{1+}(\boldsymbol{x}; \boldsymbol{d}), \ldots, u'_{n+}(\boldsymbol{x}; \boldsymbol{d}))$.

Proof: Clearly for every $\boldsymbol{x}, \boldsymbol{d} \in \mathbb{R}^s$ and $t > 0$ we obtain that

$$\begin{aligned} \frac{g(\boldsymbol{x} + t\boldsymbol{d}) - g(\boldsymbol{x})}{t} &= \\ \frac{f(\boldsymbol{u}(\boldsymbol{x} + t\boldsymbol{d})) - f(\boldsymbol{u}(\boldsymbol{x}) + t\boldsymbol{u}'_+(\boldsymbol{x}; \boldsymbol{d}))}{t} &+ \frac{f(\boldsymbol{u}(\boldsymbol{x}) + t\boldsymbol{u}'_+(\boldsymbol{x}; \boldsymbol{d})) - f(\boldsymbol{u}(\boldsymbol{x}))}{t}. \end{aligned} \tag{2.5}$$

Since f is locally Lipschitz and the Dini upper derivative of $\boldsymbol{u}$ at $\boldsymbol{x}$ in the direction $\boldsymbol{d}$ is finite it follows for $t > 0$ sufficiently small that there exists some constant $0 < L < \infty$ such that

$$\left| \frac{f(\boldsymbol{u}(\boldsymbol{x} + t\boldsymbol{d})) - f(\boldsymbol{u}(\boldsymbol{x}) + t\boldsymbol{u}'_+(\boldsymbol{x}; \boldsymbol{d}))}{t} \right| \leq L \left\| \frac{\boldsymbol{u}(\boldsymbol{x} + t\boldsymbol{d}) - \boldsymbol{u}(\boldsymbol{x})}{t} - \boldsymbol{u}'_+(\boldsymbol{x}; \boldsymbol{d}) \right\|_2$$

and this implies that

$$\limsup_{t \downarrow 0} \left| \frac{f(\boldsymbol{u}(\boldsymbol{x} + t\boldsymbol{d})) - f(\boldsymbol{u}(\boldsymbol{x}) + t\boldsymbol{u}'_+(\boldsymbol{x}; \boldsymbol{d}))}{t} \right| = 0.$$

Now the desired result follows by (2.5). □

An easy and important consequence of the above lemmas is given by the next result.

Lemma 2.5.3. *If $g : \mathbb{R}^s \longrightarrow \mathbb{R}$ is given by $g(\boldsymbol{x}) := \max(u_1(\boldsymbol{x}), \ldots, u_n(\boldsymbol{x}))$ with $u_i : \mathbb{R}^s \longrightarrow \mathbb{R}$, $i = 1, \ldots, n$, having finite Dini upper derivatives then for every $\boldsymbol{x}, \boldsymbol{d} \in \mathbb{R}^s$ it follows that*

$$g'_+(\boldsymbol{x}; \boldsymbol{d}) = \max\{u'_{i+}(\boldsymbol{x}; \boldsymbol{d}) : i \in I(\boldsymbol{x})\}$$

with $I(\boldsymbol{x}) := \{1 \leq i \leq n : g(\boldsymbol{x}) = u_i(\boldsymbol{x})\}$.

Proof: It is not difficult to verify that the function $f(x) := \max\{x_1, \ldots, x_n\}$ is locally Lipschitz and so the desired result follows by Lemma 2.5.1 and Lemma 2.5.2. □

As a final remark before concluding this chapter we mention that more results on convexity and generalizations can be found in the books by Rockafellar (Ref. [89]), Hiriart-Urruty and Lemaréchal (Ref. [52, 53]), Clarke (Ref. [19]) and Avriel, Diewert, Schaible, and Zang (Ref. [7]).

III Convex Programming

This chapter starts by introducing in Section 3.1 a *convex program*. Many important applications reduce to instances of such a problem and hence it is very important to be able to solve them. The vast majority of the existing algorithms to solve convex programs is based on so-called *separation oracles*. Among them, the *ellipsoid method* is discussed in detail in Section 3.2. In order to motivate the ellipsoid method Section 3.2 starts by considering in Subsection 3.2.1 its one-dimensional counterpart, the well-known bisection method. After a mathematical description of the ellipsoid method in Subsection 3.2.2 some improvements for the basic version of the method are introduced in Subsection 3.2.3. These improvements include a new type of *cuts*, designated by *norm cuts*, introduced by Frenk, Gromicho and Zhang in Ref. [38]. The use of norm cuts provides not only improvements of a practical nature but it also proves being a strong theoretical tool. In fact, as shown in Subsection 3.2.4, the use of these cuts permits the derivation of very simple and geometrically oriented convergence proofs. However, in practice the important question is finite termination and so Subsection 3.2.5 measures the effort taken by the method to achieve a desired accuracy. Two criteria of accuracy are considered, an absolute and a relative error, and Section 3.3 discusses stopping rules for the ellipsoid method which, if satisfied, guarantee that a given stopping criterion has been satisfied. Finally, Section 3.4 describes in general terms how the computational experiments whose results are reported in this book were conducted.

3.1. Introduction

The term *convex programming* applies to the problem of minimizing a convex function over a closed convex set. In order to derive algorithms for solving convex programs one clearly makes use of the specific properties of these optimization problems. The study of these properties belongs to the field of convex analysis and, although

Chapter 2 briefly introduces some relevant issues in this field, we refer to Rockafellar (Ref. [89]) and to Hiriart-Urruty and Lemaréchal (Ref. [52]) for completeness. The second book by Hiriart-Urruty and Lemaréchal (Ref. [53]) mostly elaborates on the design of algorithms to solve convex programming problems and also represents an important reference.

To mention some fields of quantitative decision making these problems occur frequently in engineering, economics, management science and location theory. In this book we only consider *finite dimensional* instances of the problem. Therefore, as mentioned in Chapter 2, our universe is the s-dimensional Euclidean space, $\mathbb{R}^s$. To be more precise, a finite dimensional convex optimization problem is given by

$$(P) \quad \inf\{f(x) : x \in \mathbb{R}^s,\ g_i(x) \leq 0,\ i = 1, \ldots, m\}$$

with $f, g_i : \mathbb{R}^s \longrightarrow \mathbb{R}$, $i = 1, \ldots, m$, denoting a set of convex finite-valued functions over the s-dimensional Euclidean space. Since each function g_i, $i = 1, \ldots, m$, is continuous on $\mathbb{R}^s$ (Corollary 2.3.1) it follows that the set of feasible solutions, the so-called *feasible set*, is in fact a closed convex set. In the remainder we assume that the feasible set is nonempty.

Many properties of these optimization problems are known. Among them we mention the availability of the necessary and sufficient Karush-Kuhn-Tucker optimality conditions and duality results (Ref. [52, 89, 90]). One might presume that by exploiting these properties many efficient algorithms to solve general instances of (P) would exist in the literature. However, it turns out that a large subclass of these algorithms only deals with special instances of (P). A well-known example is the simplex method for linear programming. On the other hand, efficient algorithms which can be applied to general instances of (P) assume in most cases that the *objective function* f and the *constraint functions* g_i, $i = 1, \ldots, m$, are differentiable (Ref. [26, 68]).

The algorithms discussed in this book do not require differentiability but instead they assume the existence of a separation oracle for each of the functions involved. This is clarified in the sequel. Moreover, since the maximum of a finite number of finite-valued convex functions is finite-valued and convex we may take in the definition of (P), without loss of generality, the number of different constraint functions equal to one, i.e. $m = 1$. In fact, $g_i(x) \leq 0$ for every $i = 1, \ldots, m$ if and only if $\max_{1 \leq i \leq m} g_i(x) \leq 0$. For simplicity, we denote this single constraint function by g instead of g_1. A similar argument applies to the objective function f and so optimization problem (P) also covers min-max problems. Observe that the minimization of the maximum of finitely many functions is an important, and common, criterion. Examples in the field of location theory can be found in Chapter 4 and Chapter 6.

In order to describe an algorithm to solve (P) we make the following assumption. Recall from Chapter 2 that $\mathcal{B}$ denotes the open Euclidean unit ball.

Assumption 3.1.1. *The optimization problem (P) has an optimal solution x_* satisfying $x_* \in a_0 + r\mathcal{B}$ with $a_0 \in \mathbf{R}^s$ and $r > 0$ known in advance.*

By the above observations we may assume throughout this book (without loss of generality) that a convex programming problem is defined by two convex functions, the objective function and the constraint function, and therefore it has the following description.

$$(C) \quad \min\{f(x) : x \in \mathbf{R}^s,\ g(x) \leq 0\}.$$

Hence, resorting to the notation presented in Subsection 2.3.2, the feasible set of (C) corresponds to $\mathcal{L}_g^{\leq}(0)$, i.e. the *zero lower level set* of the function g.

One class of algorithms suitable for optimizing the convex nonsmooth optimization problem (C) consists of variants of the *cutting plane* method devised independently by Kelley in Ref. [58] (see also Ref. [68]) and by Cheney and Goldstein in Ref. [18]. Cutting plane methods have been subjected to intensive research originating many interesting developments (Ref. [5, 6, 27, 47, 56, 105, 113]). A strongly related class of algorithms is given by the so-called *bundle* methods (Ref. [53]).

On the other hand, a second class of algorithms (also suitable for optimizing (C) and related to the algorithms discussed in this book) is commonly referred to as the *ellipsoid method.* This method was originated by the work of Shor (Ref. [96, 97]) and its explicit statement is also due to Shor (Ref. [98]). Moreover, Yudin and Nemirovskiĭ (Ref. [75, 114, 115]) observed its implications in convex programming.

Both classes of algorithms are based on the separation properties of subgradients and they refine in different ways the so-called subgradient method originally discovered by Shor in the early sixties and made known by Poljak (Ref. [84]). The evolution of the subgradient method into what led to the ellipsoid method is very well described by Shor in Ref. [100].

Since all these optimization methods rely on separation oracles most of the questions tackled and answered in this book are relevant to all of them. As an example we mention separation for quasiconvex functions discussed in Chapter 5. However, the ellipsoid method was chosen as a starting point for the algorithmic developments of this book for a multiplicity of reasons, perhaps the most important being its versatility. In fact, its applicability is only limited by the availability of a so-called *separation oracle*, and as one can see in Chapter 5 it can be applied to a wider class of continuous optimization problems than the extensions by Plastria (Ref. [80]) of the cutting plane methods. Observe that these extensions highly depend on subgradient-type inequalities. On the other hand, a remarkable fact about the algorithm of Plastria given in Ref. [81] is that it always exhibits feasible solutions, but in order to do so it requires the specification of an initial feasible solution.

Many methods address the question of feasibility, i.e. once specified a (in general convex) set of so-called *feasible solutions* exhibit an element of this set or prove that it is empty, see for instance Ref. [47]. The later case means in general *inconsistency* of so-called *constraints*. However, all such methods assume that the feasible set (if nonempty) is full-dimensional, i.e. has interior points. Under the same assumptions the ellipsoid method is guaranteed to exhibit feasible solutions, if they exist.

Another important characteristic of the ellipsoid method is its simplicity, which is strongly related to its versatility. Moreover, we were able to analyze its behavior in much simpler terms than the ones previously found in the literature, as well as strengthen and adapt it in a variety of ways.

Finally, although it is now acknowledged that the ellipsoid method is far from being a practical way to solve linear programming problems, we believe that it still represents a practical and competitive method in nonlinear programming. Our opinion is supported by our computational experience, carried out over several models arising in location theory, see Chapter 4 and Chapter 6. Also the experience of others, see for instance Ref. [29–31], supports this opinion.

In spite of its origins in convex non-differentiable optimization, the ellipsoid method achieved its highest point of celebrity after the work of Khachiyan (Ref. [59]). This work pioneered many recent developments in complexity theory by showing, through the use of the ellipsoid method, that linear programming belongs to the class of problems solvable in polynomial time. This settled an old open question.

The implications of the ellipsoid method to combinatorial optimization can be found in the excellent book by Grötschel, Lovász and Schrijver (Ref. [51]). In this book, using refinements of the technique pioneered by Khachiyan in Ref. [59], the polynomial time solvability of several problems is shown.

The analysis of complexity of convex programming started with the work of Yudin and Nemirovskiĭ compiled in Ref. [75]. As they do, we assume a "sufficiently exact arithmetic". By this it is meant that when dealing with issues as the effort necessary to meet some criterion of accuracy it is assumed that the values involved are computed with a precision superior to the accuracy required.

We now focus on the ellipsoid method as an optimization procedure to solve (C).

3.2. The ellipsoid method

We start this section with a result which is an immediate consequence of the subgradient inequality. This result is extremely useful in the sequel.

Lemma 3.2.1. *Let* $\varphi : \mathbf{R}^s \longrightarrow \mathbf{R}$ *be a convex function. If* $\boldsymbol{a}^\star \in \partial\varphi(\boldsymbol{a})$ *and* $h(\boldsymbol{x}) := \boldsymbol{a}^{\star\mathsf{T}}(\boldsymbol{x} - \boldsymbol{a})$ *then* $\mathcal{L}^{\leq}_{\varphi}(y) \subseteq \mathcal{L}^{\leq}_{h}(y - \varphi(\boldsymbol{a}))$ *for every* $y \in \mathbf{R}$.

Proof: From the subgradient inequality we obtain that $\varphi(\boldsymbol{x}) \geq \varphi(\boldsymbol{a}) + h(\boldsymbol{x})$ for every $\boldsymbol{x} \in \mathbf{R}^s$. Now the result follows immediately from the definition of a lower level set. □

Observe that this result can also be seen geometrically as a consequence of Theorem 2.3.6, illustrated in Figure 2.7.

Notice that $\mathcal{L}_{\bar{h}}^{=}(y)$ is a hyperplane for every $y \in \mathbf{R}$ and thus $\mathcal{L}_{\bar{h}}^{\leq}(y)$ is a halfspace.

We shall call this set, as in Chapter 2, a *lower halfspace.*

We start our analysis of the ellipsoid method by considering first the one dimensional case.

3.2.1. The one dimensional case

The ellipsoid method is easily motivated from its one dimensional counterpart. Consider first the problem of minimizing a function $f : \mathbf{R} \longrightarrow \mathbf{R}$ without the presence of constraints. Assume that an interval with finite endpoints can be constructed such that it includes an optimal solution.

The algorithm proceeds by systematically reducing this *localization interval*, always keeping the property that it includes an optimal solution, until its length meets some accuracy requirement. Once this accuracy is achieved it is assumed that any difference between the interval's midpoint and a true solution can be tolerated.

In order to realize such an algorithm we first assume the existence of an oracle which for each given point returns one of two possible answers, *greater* or *smaller*. Clearly, the answer means that optimal solutions should be found among the points greater, or smaller, than the proposed one. Since the oracle is not assumed to detect optimality, greater is interpreted as greater or equal and smaller as smaller or equal.

We mention here that an oracle is generally understood as a procedure which provides an answer to a question without any clue on how the answer is obtained. This does not mean that oracles remain obscure entities throughout this book but only that at each point where they are used the actual details of how they operate are of little or no relevance.

With these basic tools an algorithm can immediately be constructed. A possible strategy consists of applying the oracle at the midpoint of the interval and based on the answer replace the current interval by one of the two half-intervals. Taking now this new interval one can repeat the procedure as many times as necessary in order to achieve the desired accuracy. In the case of a convex function it is well-known that the desirable oracle is given by the sign of any subgradient, i.e. if a subgradient is negative the minimum is greater than or equal to the current value, and if it is positive the minimum is smaller than or equal to the current value. If it happens that the selected subgradient is zero then the current value minimizes the function.

Five iterations of this so-called *bisection method* are represented in Figure 3.1 where, without loss of generality, a differentiable function is used to illustrate the method. In this example the first four answers given by the oracle were "smaller", "smaller", "greater" and "smaller". The fifth midpoint has a null derivative and is therefore the minimum point.

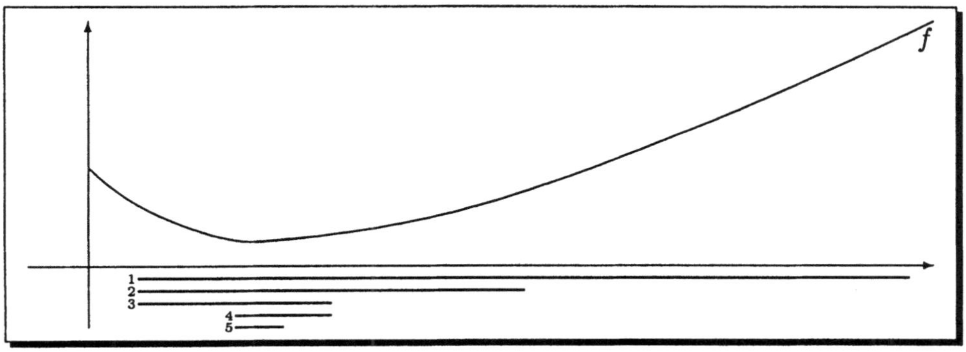

Figure 3.1: The bisection method

Algorithm 3.1 describes this basic procedure assuming that a_0 and $r_0 > 0$ are given numbers with the property that x_* belongs to $[a_0 - r_0, a_0 + r_0]$. Observe that the availability of these numbers is guaranteed by Assumption 3.1.1.

This algorithm includes two statements which are not completely defined. Such statements are marked as *framed statements.* To guarantee finite termination of the procedure we need to specify a *stopping criterion.* Examples of such criteria are provided in Section 3.3. In the present case of the bisection method it is usually reasonable to stop when the length of the localization interval has become sufficiently small. If f is convex implementing the second framed statement reduces to knowing whether f decreases or increases at a_m and this in turn amounts to knowing the sign of a subgradient at a_m.

Consider now a genuine instance of (C) for $s = 1$. The problem consists of minimizing a given convex function subject to the constraint that the solution should not make another convex function have a positive value. Such a situation is exemplified in Figure 3.2 which also shows how a constrained version of the previous algorithm works. In order to develop this algorithm we need an enhanced set of oracles.

Step 0 let $m := 0$ and $r_m := r$;
Step 1 *if* a_m satisfies some stopping criterion
then stop
else proceed to **Step 2**;
Step 2 let $r_{m+1} := r_m/2$;
if the objective function oracle says "greater" than a_m
then let $a_{m+1} := a_m + r_{m+1}$
else let $a_{m+1} := a_m - r_{m+1}$;
Step 3 let $m := m + 1$ and return to **Step 1**

Algorithm 3.1: Unconstrained basic bisection method

For each given point we would like to have an oracle telling us whether it is *feasible* or *infeasible*. If the point is feasible then we need to consult the oracle used in Algorithm 3.1. Otherwise, we need to consult a similar oracle which regards the constraint instead of the objective function.

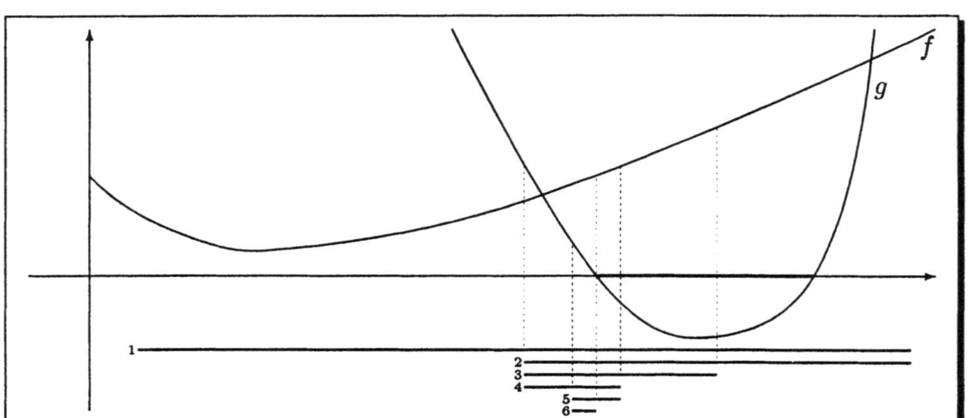

Figure 3.2: The bisection method for constrained problems

The example provided in Figure 3.2 shows an initial interval containing all the feasible solutions, and thus in particular the optimal one. The midpoint of this first interval is infeasible and so the constraint oracle is consulted. This oracle says that the feasible values can be found among those greater than the present one and interval 2 becomes the current interval. Its midpoint turns out to be feasible and this time the objective function oracle is consulted saying that smaller values should be taken. The same happens in the next iteration while the fourth midpoint becomes infeasible and once again the constraint oracle is consulted. The fifth midpoint is feasible and also optimal. However, without an oracle to certify its optimality the algorithm proceeds. In this case by consulting the objective function oracle and taking interval 6 as the next current interval we are left with only one (the optimal) feasible point in this interval. It is now easy to see that an infinite sequence (in the absence of a numerical stopping criterion) of infeasible midpoints is generated from this iteration onwards.

Step 0 let $m := 0$ and $r_m := r$;

Step 1 *if* [a_m satisfies some stopping criterion]

then stop

else proceed to **Step 2**;

Step 2 let $r_{m+1} := r_m/2$;

if a_m *is feasible*

then if [the objective function oracle says "greater" than a_m]

then let $a_{m+1} := a_m + r_{m+1}$

else let $a_{m+1} := a_m - r_{m+1}$

else if [the constraint function oracle says "greater" than a_m]

then let $a_{m+1} := a_m + r_{m+1}$

else let $a_{m+1} := a_m - r_{m+1}$;

Step 3 let $m := m + 1$ and return to **Step 1**

Algorithm 3.2: Constrained basic bisection method

Observe that this strategy guarantees finding feasible solutions in a finite number of iterations under the assumption that the part of the feasible set included in the first localization interval has full dimension, i.e. it is an interval with positive length, say $\varepsilon_1 > 0$. In fact, while each generated center is infeasible the interval of feasible solutions included in the first localization interval remains included in each current localization interval. Since the length of the first localization interval is $2r_0$ and each iteration halves the length of the current localization interval, it follows that after $\lceil \log_2(r_0/\varepsilon_1) \rceil$ iterations[1] the set of feasible solutions can not stay included in the current localization interval. Therefore, the first feasible solution must be found during the first $\lceil \log_2(r_0/\varepsilon_1) \rceil$ iterations. Furthermore, if the set of feasible solutions with lower (or equal) value than the solution found is an interval with length $\varepsilon_2 > 0$ the same reasoning applies. Thus, only when an optimal solution has been found, as in the example, it may happen that the set of feasible solutions with a lower (or equal) value than this one reduces to a point (an interval with zero length) and hence an infinite sequence of infeasible solutions starts being generated.

The previous discussion holds also in the less obvious setting of the multidimensional method presented in the next subsection, see Corollary 3.2.1 in Subsection 3.2.4. A more remarkable result, Theorem 3.3.2, shows that in that case even if an optimal solution has already been found the method does not stop generating feasible solutions.

The constrained version of this algorithm is given in Algorithm 3.2 and is just as simple as for the unconstrained case. The only difference is that **Step 2** involves more decisions and the use of more oracles.

Until now very simple oracles have been considered. These oracles only know signs of functions (e.g. the oracle deciding if a point is feasible or not) and signs of subgradients. If these oracles are able to supply actual values then an improved version of the method can be developed. In this case, it is well-known that $f(x) \geq f(a_m) + h_m(x)$ holds for every x with $h_m(x) := a_m^\star(x - a_m)$ and $a_m^\star \in \partial f(a_m)$. Applying Lemma 3.2.1 leads to $\mathcal{L}_f^{\leq}(y) \subseteq \mathcal{L}_{h_m}^{\leq}(y - f(a_m))$ for every $y \in \mathbb{R}$.

Observe that Algorithm 3.1 consists of replacing $[a_m - r_m, a_m + r_m]$ by its intersection with $\mathcal{L}_{h_m}^{\leq}(0)$ in each iteration. With this in mind, if an upper bound on the optimal function value, say $\ell \geq f(x_*)$ with x_* the minimum point, is known then by Lemma 3.2.1 it follows that $x_* \in \mathcal{L}_f^{\leq}(\ell) \subseteq \mathcal{L}_{h_m}^{\leq}(\ell - f(a_m))$. Hence one can use this set

[1]By $\lceil x \rceil$ we denote the *ceiling* of $x \in \mathbb{R}$ and by $\log_2(x)$ we denote the logarithm base 2 of $x \in \mathbb{R}_+$.

to improve the interval whenever $\ell < f(a_m)$, i.e. $\mathcal{L}^{\leq}_{\bar{h}_m}(\ell - f(a_m)) \subset \mathcal{L}^{\leq}_{\bar{h}_m}(0)$. In this case one can replace $[a_m - r_m, a_m + r_m]$ by its intersection with $\mathcal{L}^{\leq}_{\bar{h}_m}(\ell - f(a_m))$. We refer to this operation as the application of an *objective cut* to the current interval. This cut is called *deep* if $\ell < f(a_m)$ and this situation is illustrated in Figure 3.3.

In order to describe such a *deep bisection method* we need to provide a mechanism for the generation of upper bounds on the value of the optimal solution. Let us set $\ell_{-1} := +\infty$ and define for $m \geq 0$

$$\ell_m := \begin{cases} \min\{\ell_{m-1}, f(a_m)\} & \text{if } g(a_m) \leq 0 \\ \ell_{m-1} & \text{otherwise} \end{cases}.$$

One immediately sees that $\ell_m = \min\{f(a_k) : g(a_k) \leq 0,\ 0 \leq k \leq m\}$ and therefore we refer to ℓ_m as the *lowest recorded objective function value* until iteration m. This is by construction a nonincreasing sequence of upper bounds and since the sequence $f(a_m)$ with a_m feasible and $m \geq 0$ is in general not monotonous it may happen that $\ell_m < f(a_m)$. Observe that $\ell_m \leq f(a_m)$ always holds.

Clearly, as the discussion after Figure 3.2 shows, it is not only important to remember ℓ_m but also the associated *"best" recorded feasible solution*, i.e. that feasible solution a'_m with objective function value equal to ℓ_m.

It is a simple matter to verify that the new interval's midpoint is given by[2]

$$a_{m+1} := a_m - \operatorname{sign}(a^{\star}_m)\frac{r_m}{2}(1 + \alpha_m) \tag{3.1}$$

while the new interval's radius equals

$$r_{m+1} := \frac{r_m}{2}(1 - \alpha_m) \tag{3.2}$$

with $\ell_m \leq f(a_m)$ and $0 \leq \alpha_m := (f(a_m) - \ell_m)/(r_m|a^{\star}_m|) \leq 1$. The case $\ell_m = f(a_m)$ is identical to the basic bisection method ($\alpha_m = 0$).

[2]Observe that there is no loss of generality in assuming that $a^{\star}_m \neq 0$ since otherwise a_m would minimize f and therefore the algorithm would be stopped. Remember also that the so-called *sign function* $\operatorname{sign}(x)$ is defined in Example 2.3.1 of Chapter 2.

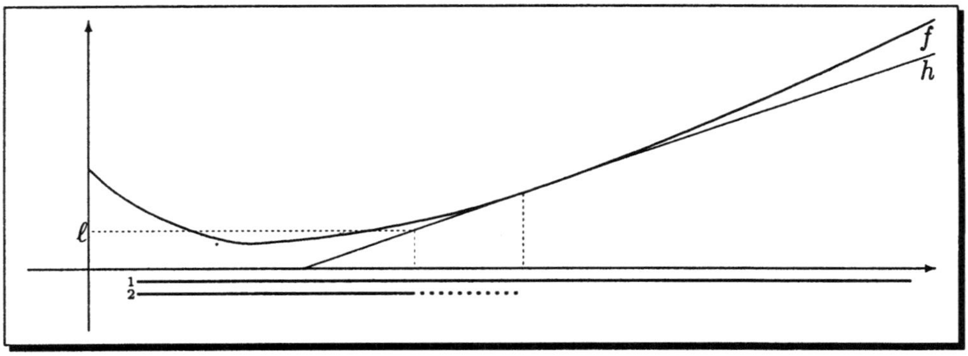

Figure 3.3: A deep objective cut in the bisection method

Having made clear how to handle a feasible point we proceed by describing what to do if a_m is infeasible. In this case $g(a_m) > 0$ and one can apply to the feasible set the same reasoning as to the lower level sets of the objective function with one advantage, namely no upper bound needs to be constructed. In fact, the feasible region is $\mathcal{L}_g^{\leq}(0)$ and by defining $h_m(x) := a_m^\star(x - a_m)$ with $a_m^\star \in \partial g(a_m)$ one concludes[3] from Lemma 3.2.1 that $\mathcal{L}_g^{\leq}(0) \subseteq \mathcal{L}_{h_m}^{\leq}(-g(a_m))$. This set can now be used to reduce the interval, see Figure 3.4. In this case we refer to such an operation as the application of a deep *constraint cut.*

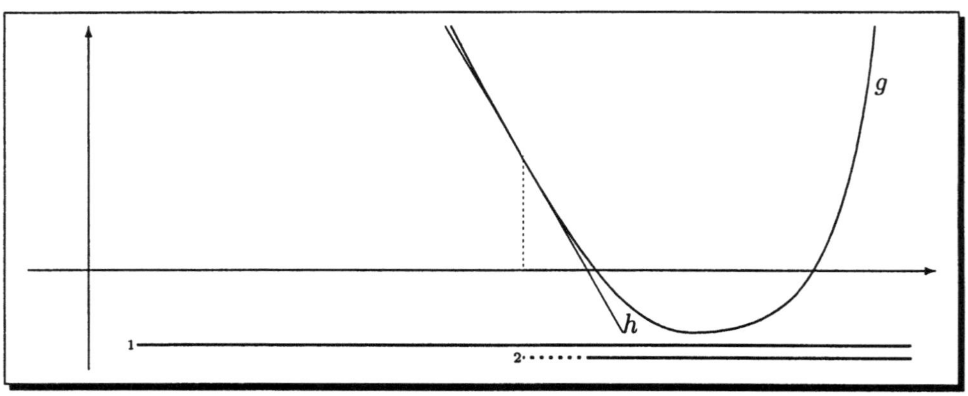

Figure 3.4: A deep constraint cut in the bisection method

[3]In this case $a_m^\star$ is necessarily not zero since we assume the existence of feasible solutions (i.e. there exists some x satisfying $g(x) \leq 0$) and $g(a_m) > 0$. Thus, a_m cannot minimize g and so $0 \notin \partial g(a_m)$ by Theorem 2.3.7.

It is quite obvious that formulas (3.1) and (3.2) still apply after taking $a_m^\star \in \partial g(a_m)$ and $\alpha_m := g(a_m)/(r_m|a_m^\star|) > 0$. The algorithm is now given by:

Step 0 let $m := 0$, $r_m := r$ and $\ell_{m-1} := +\infty$;
Step 1 *if* [a_m satisfies some stopping criterion]
  then stop
  else proceed to **Step 2**;
Step 2 *if* $g(a_m) \leq 0$
  then [apply an objective cut]
  else [apply a constraint cut];
Step 3 [update the interval], let $m := m + 1$ and return to **Step 1**

Algorithm 3.3: Deep bisection method

To complete the description of the algorithm we need to define the framed statements.

First, the application of an objective cut, which may be deep, is now specified.

if $f(a_m) < \ell_{m-1}$ *then* let $\ell_m := f(a_m)$ and $a'_m := a_m$
  else let $\ell_m := \ell_{m-1}$ and $a'_m := a'_{m-1}$;
[take $a_m^\star \in \partial f(a_m)$];
if $a_m^\star = 0$ *then* stop
  else let $\alpha_m := (f(a_m) - \ell_m)/(r_m|a_m^\star|)$

Algorithm 3.4: Application of an objective cut in the deep bisection method

Obviously the framed statement of this algorithm is function dependent. Moreover, the application of a (deep) constraint cut results into the following procedure.

$\boxed{\text{take } a_m^\star \in \partial g(a_m)}$;
let $\alpha_m := g(a_m)/(r_m|a_m^\star|)$

Algorithm 3.5: Application of a constraint cut in the deep bisection method

The framed statement of this algorithm is also function dependent.

Finally, the update of the interval in **Step 3** is accomplished by using formulas (3.1) for the center and (3.2) for the radius. This concludes the description of a *deep cut version* of the ellipsoid method for dimension one.

3.2.2. The multidimensional case

The generalization of the bisection method to more dimensions has been attempted several times. A remarkable example is the method of Wood (Ref. [112]). This so-called *multidimensional bisection* is also studied by Baritompa (Ref. [8]) who, via a dual interpretation, describes an efficient implementation for it. This method is suitable for solving global optimization problems and does not rely on separation. Moreover, when used to solve optimization problems it works in the graph space, i.e. $\mathbf{R}^{s+1}$, while our discussion concentrates on methods working in the domain space, i.e. $\mathbf{R}^s$. The method of Wood (Ref. [112]) replaces intervals by so-called *brackets* which are unions of *simplexes*. The number of simplexes in each generated bracket can grow enormously.

Levin (Ref. [65]) proposed a direct generalization (in the domain space) of the bisection method to higher dimensions resorting to separation oracles for the minimization of convex functions. His idea consisted of starting with an *interval*[4] in $\mathbf{R}^s$ which would be guaranteed to include an optimal solution x_* of (C).

The first problem is how to proceed. Contrary to the one dimensional case where the subgradient information always allows to divide an interval into two new intervals

[4]We prefer to call an interval in the s-dimensional Euclidean space a *box*, or *hyperrectangle*. Such a set can be described, similarly to its one dimensional counterpart, by just two elements of $\mathbf{R}^s$. More on boxes can be found in Chapter 5.

(for the simple reason that in $\mathbb{R}$ all the convex sets are intervals), this phenomenon does not happen in higher dimensions. Thus, after taking the center of the box and finding a separating hyperplane, one is not left with an interval but with an increasingly more complex polytope, see Figure 3.5. Consequently, not only the description of this localization set becomes progressively more complicated but the simple computation of its center soon becomes a problem whose solution requires a prohibitive amount of time. This is the second problem of Levin's method. As a result, this so-called *method of centers of gravity* is not implementable (Ref. [75]).

A remarkable result, observed by Goffin in Ref. [46] and not explicitly proved in Levin's paper, is that the polytope corresponding to the current iteration of Levin's method has its volume reduced at each iteration by a ratio of at least $1 - \left(\frac{s}{s+1}\right)^s < 1 - e^{-1}$. This implies that the minimum recorded objective function value in this method is bounded from above by a geometric progression of ratio $\sqrt[s]{1 - e^{-1}}$.

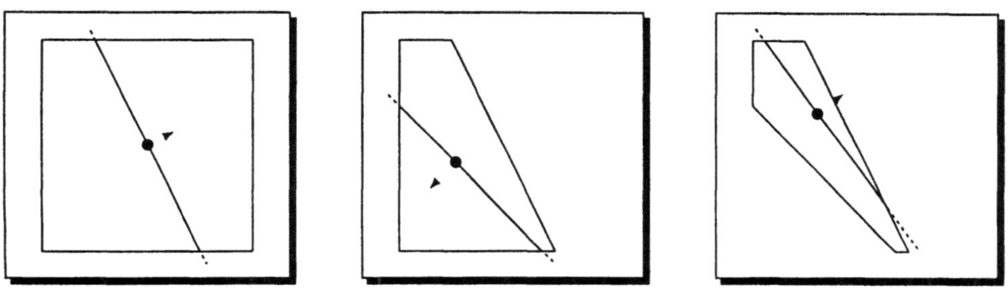

Figure 3.5: The method of centers of gravity

As the study of Yudin and Nemirovskiĭ, compiled in Ref. [75], shows no method based on separation oracles can improve significantly on the number of iterations (calls to the oracle) required by Levin's method to reach a given accuracy.

The ellipsoid method elegantly overcomes the two major problems of this method by using in each iteration a set whose description does not increase in complexity and whose center is known. Therefore this is an "ideal" generalization for an interval.

The price to pay is that instead of replacing the current localization set by one of its subsets, the ellipsoid method generates a sequence of sets whose volume decreases but each new set is not included in the previous one. Thus, each iteration of the ellipsoid method brings new points into consideration.

As Subsection 3.2.4 shows, the volume of the ellipsoids decreases at a ratio of at least $\frac{s}{s+1}\left(\frac{s^2}{s^2-1}\right)^{(s-1)/2}$ per iteration, and so this method takes approximately $2se^{-1}$ times the number of iterations (represented by calls to the oracle) that Levin's method would require to reach a given accuracy. On the other hand, the ellipsoid method is implementable.

Observe that in order to apply the ellipsoid method we need to assume the existence of an oracle that for each convex function involved in the definition of (C) is able to provide at every point of its domain the function value and an element of the subgradient set. Thus, we freely use $f(x)$ and $g(x)$ for different instances of x and take elements of the sets $\partial f(x)$ and $\partial g(x)$. One can now see the ellipsoid method as a possible generalization of Algorithm 3.3 for $s > 1$.

This approach is essentially different from Wood's (Ref. [112]) since it relies on a separation oracle. Although a disadvantage of this generalization is its dependence on a separation oracle, its advantages are, on one hand, that like traditional bisection it works on the domain space and, on the other hand, the replacements offered for intervals are simple sets which remain simple throughout the execution.

The ellipsoid method evolves as follows. Starting with a sphere (a circle in the planar case), which is an example of an ellipsoid, and cutting it through the center there exists a unique ellipsoid that contains the half ellipsoid including the optimal solution and has minimal volume among all the ellipsoids satisfying the last property. Assuming that one can compute such an ellipsoid, it may replace the initial one and the same procedure can be repeated.

As in the one dimensional case it is convenient to take less than half of each ellipsoid whenever possible. We refer to the first type of cuts, going through the center of the ellipsoid, as *central cuts* and to the second type of cuts, that leave the center of the ellipsoid in the part to be neglected, as *deep cuts*.

In Figure 3.6 one can see the difference between the two types of cuts. The first picture represents a central cut on a sphere and the minimal volume ellipsoid that includes the half sphere. The second picture shows the same sphere and a cut parallel to the first but shifted halfway along the radius of the sphere. One can see in this case that the minimum volume ellipsoid including this part of the sphere is indeed smaller than the one in the first picture.

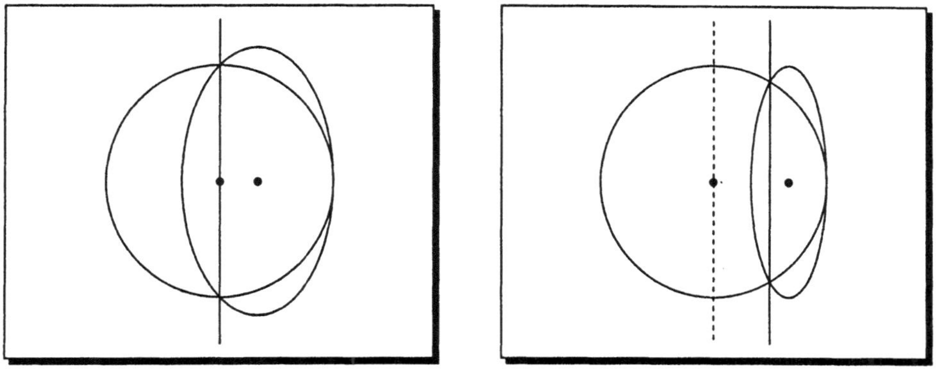

Figure 3.6: The central and the deep cut ellipsoid methods

The natural question is: how difficult is it to compute these new ellipsoids, if one finds such a (central or deep) cut? It turns out that this is very easy. In order to explain how, we first need to introduce a mathematical description of an ellipsoid.

A set $\mathcal{E} \subseteq \mathbf{R}^s$ is called an ellipsoid if there exists a vector $\boldsymbol{a} \in \mathbf{R}^s$ and a positive definite $s \times s$-matrix $\boldsymbol{A}$ such that

$$\mathcal{E} = \mathcal{E}(\boldsymbol{A};\boldsymbol{a}) := \left\{\boldsymbol{x} \in \mathbf{R}^s : (\boldsymbol{x}-\boldsymbol{a})^{\mathsf{T}}\boldsymbol{A}^{-1}(\boldsymbol{x}-\boldsymbol{a}) \leq 1\right\}.$$

Moreover, in order to determine whether a hyperplane in $\mathbf{R}^s$ with normal $\boldsymbol{a}^\star$ intersects an ellipsoid $\mathcal{E}(\boldsymbol{A};\boldsymbol{a})$ we observe (Ref. [51]) that

$$\min\{\boldsymbol{a}^{\star\mathsf{T}}\boldsymbol{x} : \boldsymbol{x} \in \mathcal{E}(\boldsymbol{A};\boldsymbol{a})\} = \boldsymbol{a}^{\star\mathsf{T}}\boldsymbol{a} - \sqrt{\boldsymbol{a}^{\star\mathsf{T}}\boldsymbol{A}\boldsymbol{a}^\star} \tag{3.3}$$

with the corresponding solution given by $\boldsymbol{a} - \boldsymbol{A}\boldsymbol{a}^\star/\sqrt{\boldsymbol{a}^{\star\mathsf{T}}\boldsymbol{A}\boldsymbol{a}^\star}$ and

$$\max\{\boldsymbol{a}^{\star\mathsf{T}}\boldsymbol{x} : \boldsymbol{x} \in \mathcal{E}(\boldsymbol{A};\boldsymbol{a})\} = \boldsymbol{a}^{\star\mathsf{T}}\boldsymbol{a} + \sqrt{\boldsymbol{a}^{\star\mathsf{T}}\boldsymbol{A}\boldsymbol{a}^\star} \tag{3.4}$$

with the corresponding solution given by $\boldsymbol{a} + \boldsymbol{A}\boldsymbol{a}^\star/\sqrt{\boldsymbol{a}^{\star\mathsf{T}}\boldsymbol{A}\boldsymbol{a}^\star}$.

This implies that the hyperplane $\mathcal{L}_h^=(\beta)$ with $h(\boldsymbol{x}) := \boldsymbol{a}^{\star\mathsf{T}}(\boldsymbol{x}-\boldsymbol{a})$ has a nonempty intersection with $\mathcal{E}(\boldsymbol{A};\boldsymbol{a})$ whenever $-\sqrt{\boldsymbol{a}^{\star\mathsf{T}}\boldsymbol{A}\boldsymbol{a}^\star} \leq -\beta \leq \sqrt{\boldsymbol{a}^{\star\mathsf{T}}\boldsymbol{A}\boldsymbol{a}^\star}$. If this holds, the hyperplane $\mathcal{L}_h^=(\beta)$ is called a *valid cut*. It is shown in Ref. [51] that for $-1/s < \alpha \leq 1$ and $\alpha := -\beta/\sqrt{\boldsymbol{a}^{\star\mathsf{T}}\boldsymbol{A}\boldsymbol{a}^\star}$ a minimum volume ellipsoid containing the intersection $\mathcal{E}(\boldsymbol{A};\boldsymbol{a}) \cap \mathcal{L}_h^{\leq}(\beta)$ has a strictly smaller volume than $\mathcal{E}(\boldsymbol{A};\boldsymbol{a})$.

To bring the exposition into the framework of an iterative algorithm let us denote the current ellipsoid by $\mathcal{E}(\boldsymbol{A}_m; \boldsymbol{a}_m)$ and the corresponding cut by $\mathcal{L}_{\overline{h}_m}^{=}(\beta_m)$ with $h_m(\boldsymbol{x}) := \boldsymbol{a}_m^{\star\mathsf{T}}(\boldsymbol{x} - \boldsymbol{a}_m)$. Finally, denoting the *depth* of the current cut by

$$0 \leq \alpha_m := \frac{-\beta_m}{\sqrt{\boldsymbol{a}_m^{\star\mathsf{T}} \boldsymbol{A}_m \boldsymbol{a}_m^{\star}}} \leq 1 \tag{3.5}$$

one can show (Ref. [14, 51]) that the ellipsoid $\mathcal{E}(\boldsymbol{A}_{m+1}; \boldsymbol{a}_{m+1})$ with center given by

$$\boldsymbol{a}_{m+1} := \boldsymbol{a}_m - \tau_m \boldsymbol{b}_m \tag{3.6}$$

and matrix

$$\boldsymbol{A}_{m+1} := \delta_m (\boldsymbol{A}_m - \sigma_m \boldsymbol{b}_m \boldsymbol{b}_m^{\mathsf{T}}) \tag{3.7}$$

with the updating values

$$\delta_m := \frac{s^2(1-\alpha_m^2)}{s^2-1}, \quad \sigma_m := \frac{2(1+s\alpha_m)}{(s+1)(1+\alpha_m)}, \quad \tau_m := \frac{1+s\alpha_m}{s+1}$$

and with $\boldsymbol{b}_m := \boldsymbol{A}_m \boldsymbol{a}_m^{\star} / \sqrt{\boldsymbol{a}_m^{\star\mathsf{T}} \boldsymbol{A}_m \boldsymbol{a}_m^{\star}}$ is the minimum volume ellipsoid containing $\mathcal{E}(\boldsymbol{A}_m; \boldsymbol{a}_m) \cap \mathcal{L}_{\overline{h}_m}^{\leq}(\beta_m)$. Observe that if $\alpha_m = 1$ then the intersection of $\mathcal{L}_{\overline{h}_m}^{\leq}(\beta_m)$ with $\mathcal{E}(\boldsymbol{A}_m; \boldsymbol{a}_m)$ reduces to the point $\boldsymbol{a}_m - \boldsymbol{b}_m$ and in this case there is no need to construct a new ellipsoid. We note here that for $\alpha_m = 0$ (respectively $0 < \alpha_m \leq 1$) the hyperplane is called a *valid central cut* (respectively *valid deep cut*).

Observe that $\boldsymbol{A}_{m+1}$ is positive definite given that $\alpha_m^2 < 1$ and that $\boldsymbol{A}_m$ is positive definite[5].

Since in this section cuts are generated by means of subgradients the following result is useful to determine when these cuts are valid.

Lemma 3.2.2. *Let $\varphi : \mathbf{R}^s \longrightarrow \mathbf{R}$ be a convex function for which a point $\widehat{\boldsymbol{x}}$ exists in $\mathcal{E}(\boldsymbol{A}; \boldsymbol{a})$ satisfying $\varphi(\widehat{\boldsymbol{x}}) \leq \varphi(\boldsymbol{a})$. Then the cut $\mathcal{L}_{\overline{h}}^{=}(\ell - \varphi(\boldsymbol{a}))$ is valid for every $\varphi(\widehat{\boldsymbol{x}}) \leq \ell \leq \varphi(\boldsymbol{a})$ with $h(\boldsymbol{x}) := \boldsymbol{a}^{\star\mathsf{T}}(\boldsymbol{x} - \boldsymbol{a})$ and $\boldsymbol{a}^{\star} \in \partial\varphi(\boldsymbol{a})$.*

[5] This is a consequence of Proposition 2.7 and Proposition 2.8 of Ref. [74].

Proof: Observe by the subgradient inequality, $\widehat{x} \in \mathcal{E}(A;a)$ and (3.3) that

$$\begin{aligned} 0 \le \varphi(a) - \ell &\le \varphi(a) - \varphi(\widehat{x}) \le a^{\star\mathsf{T}} a - a^{\star\mathsf{T}} \widehat{x} \\ &\le a^{\star\mathsf{T}} a - \min\{a^{\star\mathsf{T}} x : x \in \mathcal{E}(A;a)\} = \sqrt{a^{\star\mathsf{T}} A a^\star} \end{aligned}$$

and hence by (3.5) with $\beta := \ell - \varphi(a)$ the validity of the cut follows. □

We are now able to describe the ellipsoid method in Algorithm 3.6. This algorithm resorts to Assumption 3.1.1 to initialize a_0 and $A_0 := r^2 I$, and so $\mathcal{E}(A_0; a_0) = \mathrm{cl}(a_0 + r\mathcal{B})$. Observe the similarity between Algorithm 3.3 and Algorithm 3.6.

Step 0 let $m := 0$, $A_m := r^2 I$ and $\ell_{m-1} := +\infty$;
Step 1 *if* [a_m satisfies some stopping criterion]
then stop
else proceed to **Step 2**;
Step 2 *if* $g(a_m) \le 0$
then [apply an objective cut]
else [apply a constraint cut];
Step 3 [update the ellipsoid], let $m := m + 1$ and return to **Step 1**

Algorithm 3.6: Basic ellipsoid method

Clearly, at **Step 0** it follows by Assumption 3.1.1 that $x_* \in \mathcal{E}(A_0; a_0)$, while at step m, $m \ge 1$, it is assumed that x_* belongs to $\mathcal{E}(A_m; a_m)$.

To complete the description of the algorithm we need to specify the framed statements.

Possible stopping criteria can be found in Section 3.3.

The update of the ellipsoid in **Step 3** is done by applying formulas (3.6) to the center and (3.7) to the matrix. Moreover, to apply an objective cut we observe the following. Since f is finite-valued and convex on $\mathbf{R}^s$ it follows for every $x \in \mathcal{L}_g^{\le}(0)$ that the subgradient set $\partial f(x)$ is nonempty (Theorem 2.3.3) and hence for every $a_m^\star \in \partial f(a_m)$ the so-called subgradient inequality $f(x) \ge f(a_m) + h_m(x)$ holds with $h_m(\boldsymbol{x}) := \boldsymbol{a}_m^{\star\top}(\boldsymbol{x} - \boldsymbol{a}_m)$. Observe, if $\boldsymbol{a}_m^\star = \boldsymbol{0}$ then $\boldsymbol{a}_m$ is optimal and hence there is no need for a cut. For a derivation of a deep or central valid cut with respect to f we introduce $\ell_m := \min\{f(\boldsymbol{a}_k) : g(\boldsymbol{a}_k) \leq 0,\ 0 \leq k \leq m\}$ and observe by Lemma 3.2.1 that $\boldsymbol{x}_* \in \mathcal{L}_f^{\leq}(\ell_m) \subseteq \mathcal{L}_{h_m}^{\leq}(\ell_m - f(\boldsymbol{a}_m))$. Thus, $\boldsymbol{x}_*$ must belong to the lower halfspace $\mathcal{L}_{h_m}^{\leq}(\beta_m)$ with $\beta_m := \ell_m - f(\boldsymbol{a}_m)$. Since by assumption $\boldsymbol{x}_* \in \mathcal{E}(\boldsymbol{A}_m; \boldsymbol{a}_m)$, the validity of this cut follows by Lemma 3.2.2. Clearly, this is a valid deep cut whenever $l_m < f(\boldsymbol{a}_m)$ and this deep cut can be derived using only negligible additional computer effort. Moreover, by the observation after (3.7) we obtain that $\boldsymbol{x}_* \in \mathcal{E}(\boldsymbol{A}_{m+1}; \boldsymbol{a}_{m+1})$.

if $f(\boldsymbol{a}_m) < \ell_{m-1}$ *then* let $\ell_m := f(\boldsymbol{a}_m)$ and $\boldsymbol{a}'_m := \boldsymbol{a}_m$
else let $\ell_m := \ell_{m-1}$ and $\boldsymbol{a}'_m := \boldsymbol{a}'_{m-1}$;
take $\boldsymbol{a}_m^\star \in \partial f(\boldsymbol{a}_m)$;
if $\boldsymbol{a}_m^\star = \boldsymbol{0}$ *then* stop, $\boldsymbol{a}_m$ is optimal
else let $\alpha_m := (f(\boldsymbol{a}_m) - \ell_m)/\sqrt{\boldsymbol{a}_m^{\star\top}\boldsymbol{A}_m\boldsymbol{a}_m^\star}$;
let $\delta_m := \frac{s^2(1-\alpha_m^2)}{s^2-1}$, $\sigma_m := \frac{2(1+s\alpha_m)}{(s+1)(1+\alpha_m)}$ and $\tau_m := \frac{1+s\alpha_m}{s+1}$

Algorithm 3.7: Application of an objective cut in the ellipsoid method

Algorithm 3.7 details the application of an objective cut. Clearly, the framed statement in this algorithm is highly dependent on the function itself. Chapter 4 and Chapter 6 are largely concerned with the specification of this separation oracle for important location problems.

take $\boldsymbol{a}_m^\star \in \partial g(\boldsymbol{a}_m)$;
let $\alpha_m := g(\boldsymbol{a}_m)/\sqrt{\boldsymbol{a}_m^{\star\top}\boldsymbol{A}_m\boldsymbol{a}_m^\star}$;
let $\delta_m := \frac{s^2(1-\alpha_m^2)}{s^2-1}$, $\sigma_m := \frac{2(1+s\alpha_m)}{(s+1)(1+\alpha_m)}$ and $\tau_m := \frac{1+s\alpha_m}{s+1}$

Algorithm 3.8: Application of a constraint cut in the ellipsoid method

To apply a constraint cut we observe by Lemma 3.2.1 that $\mathcal{L}_g^{\leq}(0) \subseteq \mathcal{L}_{h_m}^{\leq}(-g(\boldsymbol{a}_m))$ with $h_m(\boldsymbol{x}) := \boldsymbol{a}_m^{\star\top}(\boldsymbol{x} - \boldsymbol{a}_m)$ for some nonzero $\boldsymbol{a}_m^\star \in \partial g(\boldsymbol{a}_m)$ and hence $\boldsymbol{x}_* \in \mathcal{L}_g^{\leq}(0) \subseteq$

$\mathcal{L}^{\leq}_{\bar{h}_m}(\beta_m)$ with $\beta_m := -g(\boldsymbol{a}_m)$. Moreover, since $\boldsymbol{x}_* \in \mathcal{E}(\boldsymbol{A}_m; \boldsymbol{a}_m)$ and by assumption $g(\boldsymbol{x}_*) \leq 0 < g(\boldsymbol{a}_m)$ the hyperplane $\mathcal{L}^{=}_{\bar{h}_m}(\beta_m)$ is a valid cut by Lemma 3.2.2. Clearly, it is a valid deep cut since we assume that $g(\boldsymbol{a}_m) > 0$. As for the case of an objective cut we obtain $\boldsymbol{x}_* \in \mathcal{E}(\boldsymbol{A}_{m+1}; \boldsymbol{a}_{m+1})$. Algorithm 3.8 details the application of a constraint cut. Notice that the framed statement in this algorithm is also function dependent.

Algorithm 3.6 is similar to the variant V1V3 of the ellipsoid algorithm studied in Ref. [29]. This technique to generate deep cuts using the subgradient inequality was first proposed by Shor and Gershovich in Ref. [101].

3.2.3. Improving the numerical stability

The ellipsoid algorithm is well-known for the occurrence of numerical instabilities. These instabilities are related to the fact that each time a new ellipsoid is generated, this new ellipsoid is not completely included in the previous one, bringing new points into consideration. As we observed in Subsection 3.2.1 this phenomenon does not occur for $s = 1$.

The inclusion of new points may induce the ellipsoid to elongate along one of its axes in such a way that eventually it may become "flat" on one of the other $s - 1$ axes. This situation creates two types of problems.

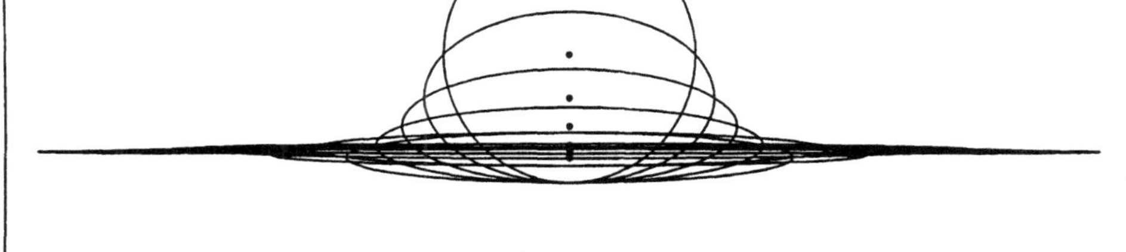

Figure 3.7: A "bad" instance for the ellipsoid method

The first problem is related to the computation of the value $\sqrt{\boldsymbol{a}_m^{\star\top} \boldsymbol{A}_m \boldsymbol{a}_m^{\star}}$. If the matrix A_m corresponds to an ellipsoid which is almost flat along one of its axes then it is almost positive semi-definite, instead of positive definite. Therefore, $\sqrt{a_m^{\star\top} A_m a_m^\star}$ may be truncated to zero when calculated by a computer for nonzero $a_m^\star$ and this induces "division by zero" in the calculation of α_m and b_m.

As the next apparently innocent example shows this situation is very likely to occur. Suppose that $s = 2$ and we take the objective function $f(x) := |x_2 + 3/4|$ with no constraint. The optimal solution value is 0 and the optimal solution set is $\mathcal{L}_f^{\leq}(0) := \{x \in \mathbb{R}^2 : x_2 = -3/4\}$. If we take then $a_0 := 0$ and $r := 1$ to construct a starting ellipsoid, then the optimal solution $x_* := (0, -3/4)$ is interior to this ellipsoid. The first 10 generated ellipsoids are represented in Figure 3.7.

Even worse, the accumulation of roundoff errors produced by the inexact computer arithmetic may lead to indefinite instances of A_m by applying (3.7) "as it is". When this happens, $a_m^{\star\top} A_m a_m^\star$ may become negative. This originates a numerical error while attempting to extract the square root.

Therefore, following Ref. [48] we implemented the algorithm using a $LDL^\top$ factorization of A (Ref. [44, 45, 48]). In the immediate sequel of this subsection we follow Ref. [48] closely.

In this factorization L is a *unit lower triangular* matrix, i.e. $L = [l_{ij}]$ with

$$l_{ij} = \begin{cases} 0 & \text{if } i < j \\ 1 & \text{if } i = j \\ \text{real number} & \text{otherwise} \end{cases}$$

and D is a *positive diagonal* matrix $D = \operatorname{diag}(d_1, \ldots, d_s)$ with $d_i > 0$ for each $i = 1, \ldots, s$.

Obtaining such a factorization for a positive definite matrix $A \in \mathbb{R}^{s \times s}$ requires in general $s^3/6$ additions and multiplications (Ref. [49]). However, if the factorization of A is known and subsequently A is submitted to a rank one update, as (3.7), updating the factorization only requires $\mathcal{O}(s^2)$ operations as shown in Ref. [44, 45, 48] and now detailed.

For $m = 0$ the decomposition is trivial. In fact $A_0 = r^2 I = L_0 D_0 L_0^\top$ with $L_0 := I$ and $D_0 := \operatorname{diag}(r^2, \ldots, r^2)$.

Let us now consider $A_m = L_m D_m L_m^\top$ and compute $A_{m+1} = \delta_m(A_m - \sigma_m b_m b_m^\top)$. We start with $\widehat{a}_m^\star := L_m^\top a_m^\star$, $\gamma_m := \widehat{a}_m^{\star\top} D_m \widehat{a}_m^\star$, $v_m := \frac{1}{\sqrt{\gamma_m}} D_m \widehat{a}_m^\star$ and $w_m := L_m v_m$. First, observe that in this way $\gamma_m = a_m^{\star\top} A_m a_m^\star$ cannot become nonpositive due to roundoff errors given that D_m has a positive diagonal. This property is preserved by the updating procedure given in Algorithm 3.9.

To update the center of the new ellipsoid we simply compute $a_{m+1} = a_m - \tau_m w_m$. The ellipsoid's matrix A_{m+1} equals $L_m\left(\delta_m(D_m - \sigma_m v_m v_m^\top)\right) L_m^\top$. Therefore, the question now is how to perform the factorization $D_m - \sigma_m v_m v_m^\top = \widetilde{L}_m \widetilde{D}_m \widetilde{L}_m^\top$. Once this is accomplished it is enough to set[6] $L_{m+1} := L_m \widetilde{L}_m$ and $D_{m+1} := \delta_m \widetilde{D}_m$.

From this point on the iterate index "m" is dropped for the sake of notational convenience.

We now discuss in detail how to perform the operations listed above which involve unit lower triangular and diagonal matrices. Let x be any vector and observe due to the nature of L that $Lx = \left[\sum_{j=1}^{s} l_{ij} x_j\right] = \left[\sum_{j=1}^{i-1} l_{ij} x_j + x_i\right]$. Moreover, due to the nature of D, we have $Dx = [d_i x_i]$. Finally, $x^\top D x = \sum_{i=1}^{s} d_i x_i^2$ making it obvious that this value cannot become negative unless some of the d_i are negative.

As in Ref. [48] we can apply the algorithm given in Section 5.2 of Ref. [45] in order to compute the positive diagonal matrix $\widetilde{D} = \operatorname{diag}\left(\widetilde{d}_1, \ldots, \widetilde{d}_s\right)$ and the unit lower triangular matrix $\widetilde{L} = \left[\widetilde{l}_{ij}\right]$ with $\widetilde{l}_{ij} = v_i \xi_j$ for each $j < i$.

$t := 1 - \sigma;$
for $j := s$ *down to* 1 *do*
  $\tau := t;$
  $t := t + \sigma v_j^2 / d_j;$
  $\widetilde{d}_j := d_j \tau / t;$
  $\xi_j := -\sigma v_j / (d_j \tau).$

Algorithm 3.9: Updating the decomposition elements

Now only the products $L\widetilde{L}$ and $\delta\widetilde{D}$ need to be computed. The second product

[6] Observe that the product of unit lower triangular matrices is unit lower triangular.

is easily obtainable and yields $\mathrm{diag}\left(\delta\widetilde{d}_1, \ldots, \delta\widetilde{d}_s\right)$. The first can be computed in a "clumsy" way in $\mathcal{O}(s^3)$ but due to the special structure of both matrices and the product form $\widetilde{l}_{ij} = v_i \xi_j$ for each $j < i$, it can be performed in only $\mathcal{O}(s^2)$ as Algorithm 3.10 shows (see also Ref. [45]).

Note that in the following algorithm only the nontrivial elements are computed; the unit diagonal and the upper triangle of zeros are not mentioned. In fact, only the $s(s-1)/2$ nontrivial elements need to be stored, making this alternative even more efficient than the direct update of the matrix $\boldsymbol{A}$ and numerically more robust. Moreover, it is assumed that the result of the product $\boldsymbol{L}\widetilde{\boldsymbol{L}}$ is stored in $\boldsymbol{L}$.

```
for i := 2 to s do
    θ := v_i;
    for j := i − 1 down to 2 do
        τ   := l_ij;
        l_ij := l_ij + θ ξ_j;
        θ   := θ + τ v_j
    l_i1 := l_i1 + θ ξ_1
```

Algorithm 3.10: Multiplying two unit lower triangular matrices

Another effect of the potential elongation of the ellipsoid is that the ellipsoid's center may end up extremely far away from the initial ellipsoid. In order to try to overcome this and thus reducing the risk of numerical instability we present the concept of *norm cuts*. Such cuts were introduced by Frenk, Gromicho and Zhang in Ref. [38] and a geometrical interpretation of them is given by the Figure 3.8.

The idea is to remember Assumption 3.1.1 at each iteration and not (as in the basic version) to use this information only to initialize the algorithm. As the next section shows the inclusion of this new cut leads to very simple convergence proofs. Thus, if it happens that the center $\boldsymbol{a}_m$ of the current ellipsoid is outside the first ellipsoid then a cut is generated using the function $n(\boldsymbol{x}) := \|\boldsymbol{x} - \boldsymbol{a}_0\|_2$. This implies by Lemma 3.2.1 and Assumption 3.1.1 that $\boldsymbol{x}_* \in \mathcal{E}(\boldsymbol{A}_0; \boldsymbol{a}_0) = \mathcal{L}_{\overline{n}}^{\leq}(r) \subseteq \mathcal{L}_{\overline{h}_m}^{\leq}(r - n(\boldsymbol{a}_m))$ with $h_m(\boldsymbol{x}) := \nabla n(\boldsymbol{a}_m)^{\mathsf{T}}(\boldsymbol{x} - \boldsymbol{a}_m)$ and $\nabla n(\boldsymbol{a}_m) = (\boldsymbol{a}_m - \boldsymbol{a}_0)/n(\boldsymbol{a}_m)$. Consequently, the optimal point $\boldsymbol{x}_*$ belongs to the lower halfspace $\mathcal{L}^{\leq}_{\bar{h}_m}(\beta_m)$ with $\beta_m := r - n(\boldsymbol{a}_m)$. The validity of this norm cut follows now from Lemma 3.2.2 and $\boldsymbol{x}_* \in \mathcal{E}(\boldsymbol{A}_m; \boldsymbol{a}_m)$ and once again we obtain that $\boldsymbol{x}_* \in \mathcal{E}(\boldsymbol{A}_{m+1}; \boldsymbol{a}_{m+1})$.

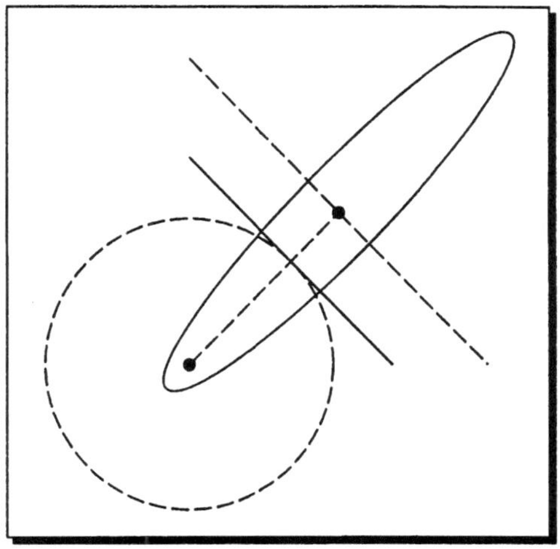

Figure 3.8: A norm cut in the ellipsoid algorithm

The improved version of the ellipsoid algorithm is described as follows.

Step 0 let $m := 0$, $\boldsymbol{A}_m := r^2\boldsymbol{I}$ and $\ell_{m-1} := +\infty$;

Step 1 *if* [$\boldsymbol{a}_m$ satisfies some stopping criterion]

then stop

else goto **Step 2**;

Step 2 *if* $n(\boldsymbol{a}_m) > r$ *then* [apply a norm cut]

else if $g(\boldsymbol{a}_m) > 0$ *then* [apply a constraint cut]

else [apply an objective cut];

Step 3 [update the ellipsoid], let $m := m + 1$ and return to **Step 1**

Algorithm 3.11: Improved ellipsoid method

The above algorithm requires the specification of four procedures, marked as framed statements. The application of a norm cut is simply done as follows.

let $\alpha_m := (n(\boldsymbol{a}_m) - r)/\sqrt{\nabla n(\boldsymbol{a}_m)^\top \boldsymbol{A}_m \nabla n(\boldsymbol{a}_m)}$;
let $\delta_m := \frac{s^2(1-\alpha_m^2)}{s^2-1}$, $\sigma_m := \frac{2(1+s\alpha_m)}{(s+1)(1+\alpha_m)}$ and $\tau_m := \frac{1+s\alpha_m}{s+1}$

Algorithm 3.12: Application of a norm cut in the improved ellipsoid method

The application of a constraint cut follows Algorithm 3.8 and the application of an objective cut follows Algorithm 3.7. The update of the ellipsoid resorts to Algorithm 3.9 and to Algorithm 3.10 in order to implement the formulas (3.6) and (3.7) in a numerically robust way. For possible stopping criteria we refer to Section 3.3. It is very important to stress here that Algorithm 3.11 slightly redefines the sequence ℓ_m generated by Algorithm 3.6. These sequences are generated by calling Algorithm 3.7 in different circumstances. The sequence generated by Algorithm 3.11 corresponds to

$$\ell_m = \min\{f(\boldsymbol{a}_k) : g(\boldsymbol{a}_k) \leq 0,\ n(\boldsymbol{a}_k) \leq r,\ 0 \leq k \leq m\}.$$

Obviously, $\boldsymbol{a}'_m$ is redefined accordingly. This subtle modification is extraordinarily important for the derivation of the convergence results included in the next subsection.

To finish this subsection observe that each unconstrained convex problem

$$(U) \quad \min\{f(\boldsymbol{x}) : \boldsymbol{x} \in \mathbf{R}^s\}$$

is equivalent to an instance of (C) by taking as a constraint any constant function satisfying $g(\boldsymbol{x}) \leq 0$ for every $\boldsymbol{x} \in \mathbf{R}^s$. However, in the sequel it is advantageous to explicitly state the unconstrained version of Algorithm 3.11. Observe that this algorithm is presented by Frenk, Gromicho and Zhang in Ref. [40] where its rate of convergence was first derived taking into account the influence of deep cuts.

Step 0 let $m := 0$, $\boldsymbol{A}_m := r^2\boldsymbol{I}$ and $\ell_{m-1} := +\infty$;
Step 1 *if* [$\boldsymbol{a}_m$ satisfies some stopping criterion]
then stop
else goto **Step 2**;
Step 2 *if* $n(\boldsymbol{a}_m) > r$ *then* [apply a norm cut]
else [apply an objective cut];
Step 3 [update the ellipsoid], let $m := m + 1$ and return to **Step 1**

Algorithm 3.13: Improved unconstrained ellipsoid method

Notice that the framed statements are implemented as in Algorithm 3.11.

This finishes our description of the ellipsoid method. In the next section we give simple and geometrically intuitive proofs of convergence for Algorithm 3.11.

3.2.4. Convergence proofs

This section provides for the sequence ℓ_m generated by Algorithm 3.11 an elementary proof of convergence to the optimal objective function value. Moreover, it also establishes the rate of convergence of this sequence in the presence of deep cuts. Goffin in Theorem 2.5 of Ref. [46] gives exactly the same convergence rate as Theorem 3.2.3 but in Ref. [46] only the central cut version applied to unconstrained problems is analyzed, and the convergence proof presented there is much more complicated. The result of Goffin corresponds to a similar result proved by Shor in Ref. [99] which seems to be the first rate of convergence proof for the ellipsoid method.

Goffin's proof was extended to the constrained case by Luthi (Ref. [69]) but still covering only central cuts. Besides, contrary to our elementary and more natural approach, a heavy result in convex analysis about volumes of concave arrays is needed in Ref. [46] and in Ref. [69]. This result can only be applied if f is convex. Observe that Frenk, Gromicho and Zhang in Ref. [38] gave the first convergence proof for Algorithm 3.11 showing the beneficial influence of deep cuts. This proof is, in opposition to the ones previously found in the literature, based on extremely simple mathematics and, as shown in Chapter 5, can be applied if f is quasiconvex. The same authors further simplified the proof in Ref. [40] but this time covering only the unconstrained case. The proof given in this subsection follows the simple techniques used in Ref. [40] and covers the constrained case. However, this proof can only be applied if f is convex and can also be found in Frenk and Gromicho (Ref. [34]).

Notice that these results do not imply the convergence of $\boldsymbol{a}_m$ to $\boldsymbol{x}_*$. The question whether this holds or not is still open. We can prove, however, that a subsequence of $\boldsymbol{a}'_m$ has the property that the minimum distance from an optimal point to it converges to zero. Also rate of convergence results at this domain level cannot be provided without much stronger assumptions, e.g. strong convexity (see Ref. [52]) of the objective function f.

We now assume that Algorithm 3.11 has already performed m iterations, $m = 1, 2, \ldots$, with centers $\boldsymbol{a}_k$, $0 \leq k \leq m$, and that no optimality check or stopping criterion was applied. We may assume, without loss of generality, that $0 \leq \alpha_k < 1$ and $\boldsymbol{a}_k^\star \neq \boldsymbol{0}$ for every $0 \leq k \leq m$ since $\alpha_k = 1$ would mean that $\mathcal{E}(\boldsymbol{A}_{k+1}; \boldsymbol{a}_{k+1})$ reduces to a point and $\boldsymbol{a}_k^\star = \boldsymbol{0}$, only possible for the objective function[7] f, would make the algorithm stop, see Algorithm 3.7.

In order to start the proof we first need the following result.

Lemma 3.2.3. *For each matrix* $\boldsymbol{A} \in \mathbb{R}^{s \times s}$ *with* $\det(\boldsymbol{A}) \neq 0$ *and each pair of vectors* $\boldsymbol{a}, \boldsymbol{b} \in \mathbb{R}^s$ *we have* $\det(\boldsymbol{A} + \boldsymbol{a}\boldsymbol{b}^\top) = (1 + \boldsymbol{b}^\top \boldsymbol{A}^{-1}\boldsymbol{a}) \det(\boldsymbol{A})$.

Proof: Observe by well-known properties of determinants (Ref. [64]) that

$$\begin{aligned} \det(\boldsymbol{A} + \boldsymbol{a}\boldsymbol{b}^\top) &= \det\left(\begin{bmatrix} 1 & \boldsymbol{b}^\top \\ 0 & \boldsymbol{A} + \boldsymbol{a}\boldsymbol{b}^\top \end{bmatrix}\right) = \det\left(\begin{bmatrix} 1 & \boldsymbol{b}^\top \\ -\boldsymbol{a} & \boldsymbol{A} \end{bmatrix}\right) \\ &= \det\left(\begin{bmatrix} 1 + \boldsymbol{b}^\top \boldsymbol{A}^{-1}\boldsymbol{a} & \boldsymbol{0}^\top \\ -\boldsymbol{a} & \boldsymbol{A} \end{bmatrix}\right) = (1 + \boldsymbol{b}^\top \boldsymbol{A}^{-1}\boldsymbol{a}) \det(\boldsymbol{A}) \end{aligned}$$

which completes the proof. □

[7] Observe that if a norm cut is applied at iteration m then $n(\boldsymbol{a}_m) > r > 0$ and since the minimum value of function n is zero, $\boldsymbol{a}_m$ cannot minimize it and hence $\boldsymbol{a}_m^\star \neq \boldsymbol{0}$ by Theorem 2.3.7. A similar argument holds for a constraint cut since in this case $g(\boldsymbol{a}_m) > 0$ and the assumption of feasibility leads to the minimum value of g (if any) being less than or equal to zero.

The following result is fundamental for our convergence proofs. For the definition of the sequences ℓ_m and β_m and the functions n and h_m we refer to the previous subsection.

Lemma 3.2.4. *For each* $m \geq 0$

$$\mathcal{L}^{\leq}_{\overline{f}}(\ell_m) \cap \mathcal{L}^{\leq}_{\overline{g}}(0) \cap \mathcal{L}^{\leq}_{\overline{n}}(r) \subseteq \mathcal{E}(\boldsymbol{A}_m; \boldsymbol{a}_m) \cap \mathcal{L}^{\leq}_{\overline{h}_m}(\beta_m) \subseteq \mathcal{E}(\boldsymbol{A}_{m+1}; \boldsymbol{a}_{m+1})$$

holds.

Proof: Recall first that $\mathcal{L}^{\leq}_{\overline{n}}(r) = \mathcal{E}(\boldsymbol{A}_0; \boldsymbol{a}_0)$. Now observe that in each iteration Algorithm 3.11 applies either an objective, a constraint or a norm cut. By Lemma 3.2.1 we obtain in case of an objective cut that $\mathcal{L}^{\leq}_{\overline{f}}(\ell_m) \subseteq \mathcal{L}^{\leq}_{\overline{h}_m}(\beta_m)$, while a constraint cut satisfies $\mathcal{L}^{\leq}_{\overline{g}}(0) \subseteq \mathcal{L}^{\leq}_{\overline{h}_m}(\beta_m)$ and a norm cut $\mathcal{L}^{\leq}_{\overline{n}}(r) \subseteq \mathcal{L}^{\leq}_{\overline{h}_m}(\beta_m)$. The result follows now easily by induction since ℓ_m is nonincreasing and we have established the validity of these cuts in Subsection 3.2.2 implying that $\mathcal{E}(\boldsymbol{A}_m; \boldsymbol{a}_m) \cap \mathcal{L}^{\leq}_{\overline{h}_m}(\beta_m) \subseteq \mathcal{E}(\boldsymbol{A}_{m+1}; \boldsymbol{a}_{m+1})$. □

In order to prove the main convergence theorem we need the following regularity condition. This condition is known as the strong Slater's condition (Ref. [52]).

Assumption 3.2.1. *There exists some* $\boldsymbol{x}_0$ *satisfying* $g(\boldsymbol{x}_0) < 0$.

Observe that this assumption together with Assumption 3.1.1 enables us to conclude that an $\widehat{\boldsymbol{x}}$ should exist such that $\widehat{\boldsymbol{x}} \in \boldsymbol{a}_0 + r\mathcal{B}$ and $g(\widehat{\boldsymbol{x}}) < 0$. In fact, by Lemma 2.3.1 we know that $\text{int}(\mathcal{L}^{\leq}_{\overline{g}}(0)) = \mathcal{L}^{<}_{g}(0)$ and since by Assumption 3.1.1 there exists a ball $\boldsymbol{x}_* + \delta\mathcal{B} \subseteq \mathcal{L}^{\leq}_{\overline{n}}(r) = \boldsymbol{a}_0 + r\mathcal{B}$ and this ball intersects $[\boldsymbol{x}_0, \boldsymbol{x}_*[\subseteq \mathcal{L}^{<}_{g}(0)$, it is enough to take $\widehat{\boldsymbol{x}}$ in this intersection.

The following result is also useful for the proof of the main convergence theorem.

Lemma 3.2.5. *Let* $\varphi : \mathbf{R}^s \longrightarrow \mathbf{R}$ *be a Lipschitz continuous function with Lipschitz constant* L_φ *and take* y *and* z *such that* $y > \varphi(z)$. *Then*

$$z + \frac{y - \varphi(z)}{L_f}\mathcal{B} \subset \mathcal{L}^{<}_{\varphi}(y).$$

Proof: Take $x \in z + \frac{y-\varphi(z)}{L_f}\mathcal{B}$. It follows that $\|x - z\|_2 < \frac{y-\varphi(z)}{L_f}$ and this implies by the Lipschitz continuity that $\varphi(x) - \varphi(z) \leq |\varphi(x) - \varphi(z)| \leq L_f\|x - z\|_2 < y - \varphi(z)$. From this inequality we obtain that $\varphi(x) < y$ and the result is proven. □

We can now prove the following convergence theorem. Observe with $a := (s^2 - 1)/s^2$ and $b := \sqrt[s]{(s+1)/(s-1)}$ that $ab > 1$ and so $\frac{1}{\sqrt{ab}} < 1$.

Theorem 3.2.1. *If the ellipsoid algorithm executes an infinite number of iterations then* $\ell_m \downarrow f(x_*)$.

Proof: We start by evaluating $\det(A_m)$. From (3.7) and Lemma 3.2.3 we obtain by induction that

$$\det(A_m) = \det(A_0) \prod_{k=0}^{m-1} \left(\delta_k^s(1 - \sigma_k)\right).$$

Since $A_0 = r^2 I$ and A_m is positive definite for every m we get after some calculations

$$0 < \det(A_m) = r^{2s} \prod_{k=0}^{m-1} \left(\frac{1}{ab}\right)^s (1 - \alpha_k^2)^s \left(\frac{1 - \alpha_k}{1 + \alpha_k}\right) \leq r^{2s} \left(\frac{1}{ab}\right)^{sm} \tag{3.8}$$

and so using $ab > 1$ it follows that $\det(A_m) \to 0$.

The convexity of $\mathcal{L}_g^<(0) \cap \mathcal{L}_n^<(r)$ together with Assumption 3.1.1 and the observation after Assumption 3.2.1 yield that the line segment $[\widehat{x}, x_*[:= \{\lambda\widehat{x} + (1-\lambda)x_* : 0 < \lambda \leq 1\}$ is contained in $\mathcal{L}_g^<(0) \cap \mathcal{L}_n^<(r)$.

It is also clear that ℓ_m is a nonincreasing sequence satisfying $\ell_m \geq f(x_*)$ for every $m \geq 0$ and so we obtain that $c := \lim_{m\uparrow\infty} \ell_m$ exists and $c \geq f(x_*)$. We now show by contradiction that c equals $f(x_*)$. Suppose therefore that $c > f(x_*)$.

Since $\mathcal{L}_f^<(c)$ is an open set including x_*, there exists an $\widetilde{x} \in [\widehat{x}, x_*[\cap \mathcal{L}_f^<(c)$. This implies using $[\widehat{x}, x_*[\subseteq \mathcal{L}_g^<(0) \cap \mathcal{L}_n^<(r)$ that $\widetilde{x}$ belongs to $\mathcal{L}_f^<(c) \cap \mathcal{L}_g^<(0) \cap \mathcal{L}_n^<(r)$. Therefore, since $\mathcal{L}_f^<(c) \cap \mathcal{L}_g^<(0) \cap \mathcal{L}_n^<(r)$ is an open set there exists some $\delta > 0$ such that $\widetilde{x} + \delta\mathcal{B} \subseteq \mathcal{L}_f^<(c) \cap \mathcal{L}_g^<(0) \cap \mathcal{L}_n^<(r)$. Hence, from Lemma 3.2.4 and $\ell_m \downarrow c$ it follows that

$$\widetilde{x} + \delta\mathcal{B} \subseteq \mathcal{L}_f^<(\ell_m) \cap \mathcal{L}_g^<(0) \cap \mathcal{L}_n^<(r) \subseteq \mathcal{E}(A_m; a_m) \cap \mathcal{L}_{h_m}^{\leq}(\beta_m).$$

Finally, using $\mathrm{vol}(\mathcal{E}(\boldsymbol{A}_m;\boldsymbol{a}_m)) = \sqrt{\det(\boldsymbol{A}_m)}v_s$ with $v_s := \mathrm{vol}(\mathcal{B})$ (Ref. [51]), we obtain from the previous inclusion that

$$\begin{aligned} 0 < \delta^s v_s = \mathrm{vol}\,(\widetilde{\boldsymbol{x}} + \delta\mathcal{B}) &\leq \mathrm{vol}\left(\mathcal{E}(\boldsymbol{A}_m;\boldsymbol{a}_m) \cap \mathcal{L}^{\leq}_{\widetilde{h}_m}(\beta_m)\right) \\ &\leq \tfrac{1}{2}\,\mathrm{vol}(\mathcal{E}(\boldsymbol{A}_m;\boldsymbol{a}_m)) = \tfrac{1}{2}\sqrt{\det(\boldsymbol{A}_m)}v_s \end{aligned}$$

for every $m \geq 0$ and this contradicts $\det(\boldsymbol{A}_m) \to 0$. Hence we must have $\ell_m \downarrow f(\boldsymbol{x}_*)$ and so the result is proven. □

Observe that the proof of Theorem 3.2.1 did not resort to the convexity of the objective function f but only to its strict lower level sets being open. This fact deserves further attention in Chapter 5.

A remarkable corollary of Theorem 3.2.1, which generalizes the discussion after Figure 3.2, is the following.

Corollary 3.2.1. *Under Assumption 3.1.1 and Assumption 3.2.1 a new feasible point which improves the current value of the sequence ℓ_m is found by Algorithm 3.11 in finitely many iterations unless $\ell_m = f(\boldsymbol{x}_*)$.*

Proof: Start by considering the initial iterates and assume that no feasible solution has been found yet. Proceeding as in the proof of Theorem 3.2.1 we have for some $\delta_1 > 0$

$$\widetilde{\boldsymbol{x}} + \delta_1\mathcal{B} \subseteq \mathcal{L}^{<}_{g}(0) \cap \mathcal{L}^{<}_{n}(r) \subseteq \mathcal{E}(\boldsymbol{A}_m;\boldsymbol{a}_m) \cap \mathcal{L}^{\leq}_{\widetilde{h}_m}(\beta_m)$$

and hence $\delta_1^s \leq \frac{1}{2}\sqrt{\det(\boldsymbol{A}_m)}$. Since after finitely many iterations this inequality ceases to hold (recall that $\det(\boldsymbol{A}_m) \downarrow 0$) one concludes that in finitely many iterations the first feasible solution is found and hence ℓ_m assumes for the first time a finite value. Suppose now that $\ell_m > f(\boldsymbol{x}_*)$. Again by the proof of Theorem 3.2.1 a $\delta_2 > 0$ exists such that

$$\widetilde{\boldsymbol{x}} + \delta_2\mathcal{B} \subseteq \mathcal{L}^{<}_{f}(\ell_m) \cap \mathcal{L}^{<}_{g}(0) \cap \mathcal{L}^{<}_{n}(r) \subseteq \mathcal{E}(\boldsymbol{A}_m;\boldsymbol{a}_m) \cap \mathcal{L}^{\leq}_{\widetilde{h}_m}(\beta_m)$$

and hence $\delta_2^s \leq \frac{1}{2}\sqrt{\det(\boldsymbol{A}_m)}$ leading to the same conclusion. Since similar reasoning applies until $\ell_m = f(\boldsymbol{x}_*)$ the result is proven. □

Observe that Corollary 3.2.1 says that it takes finitely many iterations between updates of the values of ℓ_m and not that it takes finitely many iterations for ℓ_m to reach the value $f(\boldsymbol{x}_*)$. What is guaranteed to be achieved within a finite number of iterations is the satisfaction of any prescribed *accuracy* in the result. This question is discussed in Subsection 3.2.5.

In order to analyze the behavior of the generated centers we consider the sequence $\boldsymbol{a}'_m$. By Theorem 3.2.1 this sequence is clearly well defined after a given iteration, say $\underline{m}$. Moreover, it holds by construction that $\|\boldsymbol{a}'_m - \boldsymbol{a}_0\|_2 \leq r$ for $m \geq \underline{m}$ and hence this sequence is uniformly bounded and feasible. Therefore, this sequence has an accumulation point $\boldsymbol{a}_\infty$ and there exists a subsequence $m_k \in \mathbb{N}$ satisfying $\lim_{k\uparrow\infty} \boldsymbol{a}'_{m_k} = \boldsymbol{a}_\infty$. Since f is continuous this yields $\lim_{k\uparrow\infty} f(\boldsymbol{a}'_{m_k}) = f(\boldsymbol{a}_\infty)$ and due to Theorem 3.2.1 we obtain that $f(\boldsymbol{a}_\infty) = f(\boldsymbol{x}_*)$. Moreover, since the feasible region is closed it must follow that $\boldsymbol{a}_\infty$ is feasible and so $\boldsymbol{a}_\infty$ is itself an optimal solution. Since this is true for every accumulation point of $\boldsymbol{a}'_m$ the following result holds.

Theorem 3.2.2. *If* $v_m := \min\left\{\|\boldsymbol{a}'_m - \boldsymbol{x}\|_2 : \boldsymbol{x} \in \mathcal{L}_f^{\leq}(f(\boldsymbol{x}_*)) \cap \mathcal{L}_g^{\leq}(0) \cap \mathcal{L}_n^{\leq}(r)\right\}$ *then* $\lim_{m\uparrow\infty} v_m = 0$.

After establishing the convergence of Algorithm 3.11 we proceed by deriving its rate of convergence. Recall from Corollary 2.3.1 that every finite-valued convex function on $\mathbb{R}^s$ is Lipschitz continuous on compact sets.

Theorem 3.2.3. *If f is Lipschitz continuous on $\mathcal{L}_g^{\leq}(0) \cap \mathcal{L}_n^{\leq}(r)$ with Lipschitz constant L_f and g is Lipschitz continuous on $\mathcal{L}_n^{\leq}(r)$ with constant L_g then there exists some m_0 such that*

$$0 \leq \ell_m - f(\boldsymbol{x}_*) \leq \frac{2L_f L_g r \|\boldsymbol{x}_* - \widehat{\boldsymbol{x}}\|_2}{-g(\widehat{\boldsymbol{x}})\sqrt[s]{2}} \left(\frac{1}{\sqrt{ab}}\right)^m \prod_{k=0}^{m-1} \sqrt{1-\alpha_k^2}\, \sqrt[2s]{\frac{1-\alpha_k}{1+\alpha_k}}$$

for every $m \geq m_0$.

Proof: In order to prove the stated inequality we first assume that every optimal solution $\boldsymbol{x}_*$ satisfies $g(\boldsymbol{x}_*) = 0$. Clearly, the above inequality holds if ℓ_m equals

$f(\boldsymbol{x}_*)$ for some m_0 and so we assume that $\ell_m > f(\boldsymbol{x}_*)$ holds for every $m \geq 0$. By Theorem 3.2.1 we know that $\ell_m \downarrow f(\boldsymbol{x}_*)$ and since by assumption $f(\widehat{\boldsymbol{x}}) > f(\boldsymbol{x}_*)$, there exists some m_1 such that

$$f(\boldsymbol{x}_*) < \ell_m \leq f(\widehat{\boldsymbol{x}})$$

for every $m \geq m_1$. The continuity of f enables us to create the sequence $\boldsymbol{x}_m \in [\widehat{\boldsymbol{x}}, \boldsymbol{x}_*[$ with $f(\boldsymbol{x}_m) = \ell_m$. Now, we use this sequence to create the new sequence $\widetilde{\boldsymbol{x}}_m := (\boldsymbol{x}_m + \boldsymbol{x}_*)/2$ (see Figure 3.9) and for this last sequence it follows by the convexity of f that $f(\widetilde{\boldsymbol{x}}_m) < \ell_m$.

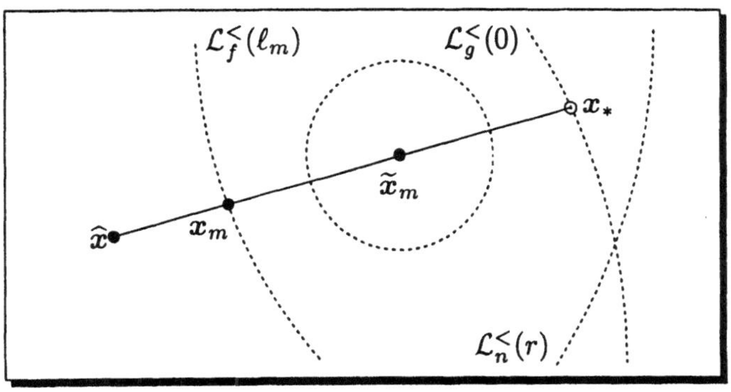

Figure 3.9: Geometric interpretation of the proof

Hence, by Lemma 3.2.5 we obtain that

$$\widetilde{\boldsymbol{x}}_m + \frac{\ell_m - f(\widetilde{\boldsymbol{x}}_m)}{L_f}\mathcal{B} \subset \mathcal{L}_f^<(\ell_m). \tag{3.9}$$

Recall now from the convexity of f and $\ell_m = f(\boldsymbol{x}_m)$ that

$$\frac{\ell_m - f(\widetilde{\boldsymbol{x}}_m)}{\|\boldsymbol{x}_m - \widetilde{\boldsymbol{x}}_m\|_2} \geq \frac{\ell_m - f(\boldsymbol{x}_*)}{\|\boldsymbol{x}_m - \boldsymbol{x}_*\|_2}$$

and, by construction, $\|\boldsymbol{x}_m - \boldsymbol{x}_*\|_2 = 2\|\boldsymbol{x}_m - \widetilde{\boldsymbol{x}}_m\|_2$. This yields that $\ell_m - f(\widetilde{\boldsymbol{x}}_m) \geq (\ell_m - f(\boldsymbol{x}_*))/2$ and thus (3.9) implies that

$$\widetilde{\boldsymbol{x}}_m + \frac{\ell_m - f(\boldsymbol{x}_*)}{2L_f}\mathcal{B} \subset \mathcal{L}_f^<(\ell_m). \tag{3.10}$$

On the other hand, by the convexity of g we obtain that $g(\widetilde{\boldsymbol{x}}_m) < 0$ and applying again Lemma 3.2.5 yields

$$\widetilde{\boldsymbol{x}}_m + \frac{0 - g(\widetilde{\boldsymbol{x}}_m)}{L_g}\mathcal{B} \subset \mathcal{L}_g^{<}(0). \tag{3.11}$$

Now, from the convexity of g, $g(\boldsymbol{x}_*) = 0$, and the Lipschitz continuity of f with Lipschitz constant L_f it follows that

$$-g(\widetilde{\boldsymbol{x}}_m) \geq -\frac{1}{2}g(\boldsymbol{x}_m) \geq -g(\widehat{\boldsymbol{x}})\frac{\|\boldsymbol{x}_* - \boldsymbol{x}_m\|_2}{2\|\boldsymbol{x}_* - \widehat{\boldsymbol{x}}\|_2} \geq -g(\widehat{\boldsymbol{x}})\frac{\ell_m - f(\boldsymbol{x}_*)}{2L_f\|\boldsymbol{x}_* - \widehat{\boldsymbol{x}}\|_2}$$

and this, together with (3.11), leads to

$$\widetilde{\boldsymbol{x}}_m + \frac{-g(\widehat{\boldsymbol{x}})(\ell_m - f(\boldsymbol{x}_*))}{2L_fL_g\|\boldsymbol{x}_* - \widehat{\boldsymbol{x}}\|_2}\mathcal{B} \subset \mathcal{L}_g^{<}(0). \tag{3.12}$$

Combining (3.10) with (3.12) and observing that $-g(\widehat{\boldsymbol{x}}) \leq L_g\|\boldsymbol{x}_* - \widehat{\boldsymbol{x}}\|_2$ finally yields

$$\widetilde{\boldsymbol{x}}_m + \frac{-g(\widehat{\boldsymbol{x}})(\ell_m - f(\boldsymbol{x}_*))}{2L_fL_g\|\boldsymbol{x}_* - \widehat{\boldsymbol{x}}\|_2}\mathcal{B} \subset \mathcal{L}_f^{<}(\ell_m) \cap \mathcal{L}_g^{<}(0).$$

Since $[\widehat{\boldsymbol{x}}, \boldsymbol{x}_*[\, \subset \mathcal{L}_n^{<}(r)$, there exists an $\varepsilon > 0$ such that $\boldsymbol{x} + \varepsilon\mathcal{B} \subset \mathcal{L}_n^{<}(r)$ for every $\boldsymbol{x} \in [\widehat{\boldsymbol{x}}, \boldsymbol{x}_*[$. Taking now m_2 such that for $m > m_2$ we have

$$\frac{-g(\widehat{\boldsymbol{x}})(\ell_m - f(\boldsymbol{x}_*))}{2L_fL_g\|\boldsymbol{x}_* - \widehat{\boldsymbol{x}}\|_2} < \varepsilon$$

and so it follows by Lemma 3.2.4 for $m > m_0 := \max\{m_1, m_2\}$ that

$$\begin{aligned}\widetilde{\boldsymbol{x}}_m + \frac{-g(\widehat{\boldsymbol{x}})(\ell_m - f(\boldsymbol{x}_*))}{2L_fL_g\|\boldsymbol{x}_* - \widehat{\boldsymbol{x}}\|_2}\mathcal{B} &\subset \mathcal{L}_f^{<}(\ell_m) \cap \mathcal{L}_g^{<}(0) \cap \mathcal{L}_n^{<}(r) \\ &\subseteq \mathcal{E}(\boldsymbol{A}_m; \boldsymbol{a}_m) \cap \mathcal{L}_{\widehat{h}_m}^{\leq}(\beta_m).\end{aligned}$$

Thus,

$$\begin{aligned}\operatorname{vol}\left(\widetilde{\boldsymbol{x}}_m + \frac{-g(\widehat{\boldsymbol{x}})(\ell_m - f(\boldsymbol{x}_*))}{2L_fL_g\|\boldsymbol{x}_* - \widehat{\boldsymbol{x}}\|_2}\mathcal{B}\right) &\leq \operatorname{vol}(\mathcal{E}(\boldsymbol{A}_m; \boldsymbol{a}_m) \cap \mathcal{L}_{\widehat{h}_m}^{\leq}(\beta_m)) \\ &\leq \tfrac{1}{2}\operatorname{vol}(\mathcal{E}(\boldsymbol{A}_m; \boldsymbol{a}_m)) = \tfrac{1}{2}\sqrt{\det(\boldsymbol{A}_m)}v_s\end{aligned}$$

and computing these volumes yields by (3.8) that

$$\left(\frac{-g(\widehat{x})(\ell_m - f(x_*))}{2L_fL_g\|x_* - \widehat{x}\|_2}\right)^s v_s \leq \frac{1}{2}\sqrt{r^{2s}\prod_{k=0}^{m-1}\left(\frac{1}{ab}\right)^s(1-\alpha_k^2)^s\left(\frac{1-\alpha_k}{1+\alpha_k}\right)}v_s.$$

Dividing the previous inequality by v_s, raising both sides to s^{-1} and multiplying by $-(2L_fL_g\|x_* - \widehat{x}\|_2)/g(\widehat{x})$ implies the desired result. To complete the proof we still have to consider the case when $g(x_*) < 0$ for some optimal solution x_*. It is not difficult to verify that for this case the same result holds, since in fact $\widetilde{x}_m$ can be taken equal to x_* for every $m > m_0$ for which $\frac{\ell_m - f(x_*)}{L_f} < \varepsilon$ with $\varepsilon > 0$ satisfying $x_* + \varepsilon\mathcal{B} \subseteq \mathcal{L}_g^<(0) \cap \mathcal{L}_n^<(r)$. Thus, from Lemma 3.2.5 we conclude that

$$x_* + \frac{\ell_m - f(x_*)}{L_f} \subseteq \mathcal{L}_f^<(\ell_m) \cap \mathcal{L}_n^<(r) \cap \mathcal{L}_g^<(0)$$

and so by a similar volume argument as used in the first part we may proceed to achieve the same result. □

Observe that Theorem 3.2.3 simultaneously establishes the rate of convergence of Algorithm 3.11 and Algorithm 3.13. However, if an instance of (U) is to be solved it follows from the last part of the proof (see also Ref. [40]) that

$$0 \leq \ell_m - f(x_*) \leq \frac{L_f r}{\sqrt[s]{2}}\left(\frac{1}{\sqrt{ab}}\right)^m \prod_{k=0}^{m-1}\sqrt{1-\alpha_k^2}\sqrt[2s]{\frac{1-\alpha_k}{1+\alpha_k}}.$$

Hence we conclude that the ellipsoid algorithm may solve instances of (U) faster than instances of (C). This speculation is supported by our computational experience reported in Chapter 4.

3.2.5. Complexity

This section assumes the specification of a value $\varepsilon > 0$ such that Algorithm 3.11 should only be executed until $\ell_m - f(x_*) \leq \varepsilon$.

Theorem 3.2.4. *Algorithm 3.11 meets the required accuracy within $\mathcal{O}(s^2\log(1/\varepsilon))$ iterations.*

Proof: From Theorem 3.2.3 we have that

$$\ell_m - f(\boldsymbol{x}_*) \leq K \left(\frac{1}{\sqrt{ab}} \right)^m$$

for some constant $K > 0$. Since $ab > e^{1/s^2}$ this yields $\ell_m - f(\boldsymbol{x}_*) \leq Ke^{-m/(2s^2)}$. The result follows now by determining the smallest value of m satisfying $e^{-m/(2s^2)} \leq \varepsilon/K$ or equivalently $m \geq 2s^2 \log(1/\varepsilon) + 2s^2 \log(K)$. □

Recall now that by Subsection 3.2.3 updating the ellipsoid in each iteration requires $\mathcal{O}(s^2)$ arithmetic operations. If A is an upper bound on the number of arithmetic operations required by each of the oracles involved then the following result is immediately clear.

Corollary 3.2.2. *Algorithm 3.11 meets the required accuracy using $\mathcal{O}(s^4 \log(1/\varepsilon) + s^2 A \log(1/\varepsilon))$ arithmetic operations.*

This concludes our theoretical analysis of the ellipsoid method.

In the next section we discuss possible stopping criteria.

3.3. Stopping criteria

In this subsection we analyze stopping criteria for Algorithm 3.11 in order to guarantee a predefined accuracy of the result with respect to the objective function value.

Two possible criteria are, once given $\varepsilon > 0$, to stop if

$$\ell_m - f(\boldsymbol{x}_*) < \varepsilon$$

or, if additionally it is known that $f(\boldsymbol{x}_*) > 0$, one may prefer to stop if

$$\frac{\ell_m - f(\boldsymbol{x}_*)}{f(\boldsymbol{x}_*)} < \varepsilon.$$

The first criterion specifies ε as the maximum allowed *absolute error* of the objective function value while the second criterion specifies ε as the maximum allowed *relative error* on the objective function value (Ref. [24]).

We present now a pair of rules (Ref. [38]) which, once satisfied, imply the satisfaction of each of these criterion.

Observe by the subgradient inequality, $\boldsymbol{x}_* \in \mathcal{E}(\boldsymbol{A}_m; \boldsymbol{a}_m)$ and (3.3) that as in the proof of Lemma 3.2.2

$$\begin{aligned} 0 \leq f(\boldsymbol{a}_m) - \ell_m &\leq f(\boldsymbol{a}_m) - f(\boldsymbol{x}_*) \leq \boldsymbol{a}_m^{\star\mathsf{T}} \boldsymbol{a}_m - \boldsymbol{a}_m^{\star\mathsf{T}} \boldsymbol{x}_* \\ &\leq \boldsymbol{a}_m^{\star\mathsf{T}} \boldsymbol{a}_m - \min\{\boldsymbol{a}_m^{\star\mathsf{T}} \boldsymbol{x} : \boldsymbol{x} \in \mathcal{E}(\boldsymbol{A}_m; \boldsymbol{a}_m)\} \\ &= \sqrt{\boldsymbol{a}_m^{\star\mathsf{T}} \boldsymbol{A}_m \boldsymbol{a}_m^{\star}}. \end{aligned} \tag{3.13}$$

By the definition of α_m this yields that the inequality

$$0 \leq l_m - f(\boldsymbol{x}_*) = l_m - f(\boldsymbol{a}_m) + f(\boldsymbol{a}_m) - f(\boldsymbol{x}_*) \leq (1 - \alpha_m)\sqrt{\boldsymbol{a}_m^{\star\mathsf{T}} \boldsymbol{A}_m \boldsymbol{a}_m^{\star}} \tag{3.14}$$

holds for every objective cut $m \geq 0$ and hence

$$l_m - f(\boldsymbol{x}_*) \leq \min\left\{(1 - \alpha_k)\sqrt{\boldsymbol{a}_k^{\star\mathsf{T}} \boldsymbol{A}_k \boldsymbol{a}_k^{\star}} : k \leq m,\ k \text{ is an objective cut}\right\}. \tag{3.15}$$

Therefore, if we are interested in an absolute error of less than a given $\varepsilon > 0$ by (3.14) this is achieved if at iteration m an objective cut is performed and the inequality

$$(1 - \alpha_m)\sqrt{\boldsymbol{a}_m^{\star\mathsf{T}} \boldsymbol{A}_m \boldsymbol{a}_m^{\star}} < \varepsilon$$

holds. However, if we know additionally that $f(\boldsymbol{x}_*) > 0$ and we are interested in the relative error then the algorithm is stopped at the mth iteration if an objective cut is performed at this iteration and the inequality

$$\begin{aligned} &\min\left\{(1 - \alpha_k)\sqrt{\boldsymbol{a}_k^{\star\mathsf{T}} \boldsymbol{A}_k \boldsymbol{a}_k^{\star}} : k \leq m,\ k \text{ is an objective cut}\right\} < \\ &\qquad \varepsilon \max\left\{f(\boldsymbol{a}_k) - \sqrt{\boldsymbol{a}_k^{\star\mathsf{T}} \boldsymbol{A}_k \boldsymbol{a}_k^{\star}} : k \leq m,\ k \text{ is an objective cut}\right\} \end{aligned}$$

holds. Observe by (3.13) that $f(\boldsymbol{x}_*) \geq f(\boldsymbol{a}_k) - \sqrt{\boldsymbol{a}_k^{\star\mathsf{T}} \boldsymbol{A}_k \boldsymbol{a}_k^{\star}}$ holds for every objective cut $k \leq m$. If the above stopping criterion is satisfied then additionally the following inequality holds

$$f(\boldsymbol{x}_*) \geq \max\left\{ f(\boldsymbol{a}_k) - \sqrt{\boldsymbol{a}_k^{\star\mathsf{T}} \boldsymbol{A}_k \boldsymbol{a}_k^\star} : k \leq m,\ k \text{ is an objective cut} \right\} > 0$$

and hence we finally obtain by (3.15) and the stopping criterion that

$$\frac{l_m - f(\boldsymbol{x}_*)}{f(\boldsymbol{x}_*)} < \varepsilon.$$

This means that we have found a feasible solution within a relative error of ε of the optimal value, or with value not exceeding $1 + \varepsilon$ times the optimal value.

The following subsection analyses under which conditions these rules are eventually satisfied.

3.3.1. Satisfaction of the stopping rules

Before analyzing the satisfaction of such conditions by the ellipsoid method we need the following well-known result from linear algebra.

Lemma 3.3.1. *For every nonsingular matrix $\boldsymbol{A} \in \mathbb{R}^{s \times s}$ and vectors $\boldsymbol{a}, \boldsymbol{b} \in \mathbb{R}^s$ such that $\boldsymbol{A} + \boldsymbol{a}\boldsymbol{b}^\mathsf{T}$ is nonsingular it holds*

$$(\boldsymbol{A} + \boldsymbol{a}\boldsymbol{b}^\mathsf{T})^{-1} = \boldsymbol{A}^{-1} - \frac{\boldsymbol{A}^{-1}\boldsymbol{a}\boldsymbol{b}^\mathsf{T}\boldsymbol{A}^{-1}}{1 + \boldsymbol{b}^\mathsf{T}\boldsymbol{A}^{-1}\boldsymbol{a}}.$$

The previous formula is called the Sherman-Morrison formula and its proof can be found in Ref. [64].

In order to prove the next lemma we introduce for every positive definite matrix $\boldsymbol{A}_k$ and $\boldsymbol{a}_k^\star \neq \boldsymbol{0}$ the ratio $\varepsilon_k := \sqrt{\boldsymbol{a}_k^{\star\mathsf{T}} \boldsymbol{A}_k \boldsymbol{a}_k^\star} / \|\boldsymbol{a}_k^\star\|_2$.

Lemma 3.3.2. *If the described algorithm has performed m steps without applying the stopping rule or an optimality check then*

$$\sum_{k=0}^{m} \left(\prod_{j=0}^{k-1} \delta_j \right) \frac{\sigma_k}{1 - \sigma_k} \varepsilon_k^{-2} \geq \frac{s}{r^2} \left(\prod_{k=0}^{m} \frac{1}{\sqrt[s]{1 - \sigma_k}} - 1 \right)$$

with $\prod_{j=0}^{-1} \delta_j = 1$.

Proof: From Lemma 3.2.3, Lemma 3.3.1 and the remarks at the beginning of Subsection 3.2.4 it is easy to verify that

$$\begin{aligned} \boldsymbol{A}_{m+1}^{-1} &= \delta_m^{-1}\left(\boldsymbol{A}_m^{-1} + \frac{\sigma_m}{1-\sigma_m}\frac{\boldsymbol{a}_m^\star \boldsymbol{a}_m^{\star\mathsf{T}}}{\boldsymbol{a}_m^{\star\mathsf{T}}\boldsymbol{A}_m\boldsymbol{a}_m^\star}\right), &(3.16)\\ \det(\boldsymbol{A}_{m+1}) &= \delta_m^s(1-\sigma_m)\det(\boldsymbol{A}_m). &(3.17)\end{aligned}$$

Moreover, if $\text{tr}(\boldsymbol{A})$ denotes the trace of matrix $\boldsymbol{A}$ this yields by (3.16) that

$$\text{tr}(\boldsymbol{A}_{m+1}^{-1}) = \delta_m^{-1}\text{tr}(\boldsymbol{A}_m^{-1}) + \delta_m^{-1}\frac{\sigma_m}{1-\sigma_m}\varepsilon_m^{-2}.$$

Iterating the previous formula for $m \geq 0$ we obtain

$$\text{tr}(\boldsymbol{A}_{m+1}^{-1}) = \text{tr}(\boldsymbol{A}_0^{-1})\prod_{k=0}^{m}\delta_k^{-1} + \sum_{k=0}^{m}\left(\prod_{j=k}^{m}\delta_j^{-1}\right)\frac{\sigma_k}{1-\sigma_k}\varepsilon_k^{-2}$$

and, since $\text{tr}(\boldsymbol{A}_0^{-1}) = \frac{s}{r^2}$, this simplifies to

$$\text{tr}(\boldsymbol{A}_{m+1}^{-1}) = \frac{s}{r^2}\prod_{k=0}^{m}\delta_k^{-1} + \sum_{k=0}^{m}\left(\prod_{j=k}^{m}\delta_j^{-1}\right)\frac{\sigma_k}{1-\sigma_k}\varepsilon_k^{-2}. \tag{3.18}$$

If (3.17) is also iterated for $m \geq 0$ it follows that

$$\det(\boldsymbol{A}_{m+1}) = \det(\boldsymbol{A}_0)\prod_{k=0}^{m}(\delta_k^s(1-\sigma_k))$$

and this implies using $\det(\boldsymbol{A}_0) = r^2 s$ and $\det(\boldsymbol{A}^{-1}) = \det(\boldsymbol{A})^{-1}$ that

$$\det(\boldsymbol{A}_{m+1}^{-1}) = r^{-2s}\prod_{k=0}^{m}(\delta_k^{-s}(1-\sigma_k)^{-1}). \tag{3.19}$$

Since $\text{tr}(\boldsymbol{A}_{m+1}^{-1}) = \sum_{i=1}^{s}\lambda_i$ and $\det(\boldsymbol{A}_{m+1}^{-1}) = \prod_{i=1}^{s}\lambda_i$ with λ_i, $i = 1, \ldots, s$, the positive eigenvalues of the positive definite matrix $\boldsymbol{A}_{m+1}^{-1}$, we obtain by the geometric-arithmetic mean inequality (Ref. [106]) that

$$s\sqrt[s]{\det(\boldsymbol{A}_{m+1}^{-1})} \leq \text{tr}(\boldsymbol{A}_{m+1}^{-1}). \tag{3.20}$$

Substituting (3.18) and (3.19) into (3.20) we finally obtain

$$\sum_{k=0}^{m}\left(\prod_{j=k}^{m}\delta_j^{-1}\right)\frac{\sigma_k}{1-\sigma_k}\varepsilon_k^{-2} \geq \frac{s}{r^2}\left(\prod_{k=0}^{m}\delta_k^{-1}\right)\left(\prod_{k=0}^{m}\frac{1}{\sqrt[s]{1-\sigma_k}}-1\right)$$

and multiplying the last expression by $\prod_{k=0}^{m}\delta_k$ the desired result follows. □

Before introducing the convergence result we also need to define the sequence $\omega_m := \min\{(1-\alpha_k)\varepsilon_k : k \leq m\}$. Recall that $a := (s^2-1)/s^2$ and $b := \sqrt[s]{(s+1)/(s-1)}$.

Theorem 3.3.1. *If the deep cut ellipsoid algorithm, without applying a stopping rule or optimality check, is executed an infinite number of iterations then*

$$\lim_{m\to\infty}\omega_m = 0.$$

Moreover, the convergence is geometric at a rate of $1/\sqrt{ab}$ *if* $\alpha_m = 0$ *for every* m *(central cut version) and at least as fast whenever* $\alpha_m > 0$ *for some* m *(deep cut version).*

Proof: The sequence ω_m is clearly nonincreasing and nonnegative. Also, without loss of generality, we may assume as observed at the beginning of Subsection 3.2.4, that $0 \leq \alpha_k < 1$ and $\boldsymbol{a}_k^\star \neq \boldsymbol{0}$ for every $k \geq 0$. Observe that after some rewriting the inequality

$$\frac{2}{s-1}\sum_{k=0}^{m}a^{-k}\prod_{j=0}^{k-1}(1-\alpha_j^2)\frac{1+s\alpha_k}{(1-\alpha_k)\varepsilon_k^2} \geq \frac{s}{r^2}\left(b^{m+1}\prod_{k=0}^{m}\sqrt[s]{\frac{1+\alpha_k}{1-\alpha_k}}-1\right)$$

follows from Lemma 3.3.2.

Since $0 \leq \alpha_k < 1$ for every k, the inequalities $\prod_{j=0}^{k-1}(1-\alpha_j^2) \leq 1$ and $1+s\alpha_k < 1+s$ are satisfied and hence the above expression implies

$$2b^s\sum_{k=0}^{m}a^{-k}\frac{1}{(1-\alpha_k)\varepsilon_k^2} > \frac{s}{r^2}\left(b^{m+1}\prod_{k=0}^{m}\sqrt[s]{\frac{1+\alpha_k}{1-\alpha_k}}-1\right). \tag{3.21}$$

By definition, $\omega_m \leq (1-\alpha_k)\varepsilon_k$ for every $k \leq m$ and since $(1-\alpha_k)^2 \leq (1-\alpha_k)$ it follows that

$$\omega_m^2 \leq (1-\alpha_k)\varepsilon_k^2.$$

Hence, by (3.21) we obtain that

$$2b^s\omega_m^{-2}\sum_{k=0}^{m} a^{-k} > \frac{s}{r^2}\left(b^{m+1}\prod_{k=0}^{m}\sqrt[s]{\frac{1+\alpha_k}{1-\alpha_k}}-1\right).$$

Now, observing that $\sum_{k=0}^{m} a^{-k} = (1-a^{-(m+1)})/(1-a^{-1})$, the last inequality yields

$$\omega_m^2 < \theta_m^2 := 2r^2\frac{(s+1)^2}{s}\frac{\left(a^{-1}\right)^{m+1}-1}{b^{m+1}\prod_{k=0}^{m}\sqrt[s]{\frac{1+\alpha_k}{1-\alpha_k}}-1}.$$

Note that $\prod_{k=0}^{m}\sqrt[s]{(1+\alpha_k)/(1-\alpha_k)} \geq 1$ for every m and so

$$\theta_m^2 \leq \tau_m^2 := 2r^2\frac{(s+1)^2}{s}\frac{\left(a^{-1}\right)^{m+1}-1}{b^{m+1}-1}.$$

Hence by using $1/(ab) < 1$ it follows that $\tau_m^2 \to 0$ as $m \to \infty$ and its geometric convergence rate is of the order $1/(ab)$. However, if $\alpha_k > 0$ for some k then θ_m^2 might decrease faster, and so this might also hold for ω_m^2. Finally, if $\omega_m^2 \to 0$ at a rate of at least $1/(ab)$, then $\omega_m \to 0$ at a rate of at least $1/\sqrt{ab}$. □

It remains to relate the behavior of the sequence ω_m, $m \geq 0$, to the behavior of the nonincreasing sequence $l_m - f(x_*)$, $m \geq 0$. In order to do so, the following notation is necessary.

$$\begin{aligned}
\omega_m^n &:= \min\{(1-\alpha_k)\varepsilon_k : k \leq m,\ k \text{ is a norm cut}\},\\
\omega_m^g &:= \min\{(1-\alpha_k)\varepsilon_k : k \leq m,\ k \text{ is a constraint cut}\},\\
\omega_m^f &:= \min\{(1-\alpha_k)\varepsilon_k : k \leq m,\ k \text{ is an objective cut}\}.
\end{aligned}$$

To avoid ambiguities, ω_m^n, ω_m^g and ω_m^f are set equal to infinity if the corresponding sets are empty. Since at each step the algorithm either performs a norm cut, a constraint cut or an objective cut, it is clear that $\omega_m = \min\{\omega_m^n, \omega_m^g, \omega_m^f\}$.

It is now possible to prove the next result for ω_m^n.

Lemma 3.3.3. *There exists a positive constant $\delta > 0$ such that $\omega_m^n \geq \delta$ for every $m \geq 0$.*

Proof: Let $m \geq 0$ be given and suppose there exists some $k \leq m$ such that the algorithm performs a norm cut during step k. If such a k does not exist the proof is finished since in that case $\omega_m^n = +\infty$. To continue the analysis of the first case we observe by Assumption 3.1.1 that there exists some $\delta > 0$ such that $n(\boldsymbol{x}_*) \leq r - \delta$ and hence by the subgradient inequality and (3.3) it follows that

$$\begin{aligned}
r - \delta &\geq n(\boldsymbol{x}_*) \\
&\geq n(\boldsymbol{a}_m) + \nabla n(\boldsymbol{a}_m)^\intercal(\boldsymbol{x}_* - \boldsymbol{a}_m) \\
&= n(\boldsymbol{a}_m) - \nabla n(\boldsymbol{a}_m)^\intercal \boldsymbol{a}_m + \nabla n(\boldsymbol{a}_m)^\intercal \boldsymbol{x}_* \\
&\geq n(\boldsymbol{a}_m) - \nabla n(\boldsymbol{a}_m)^\intercal \boldsymbol{a}_m + \min\{\nabla n(\boldsymbol{a}_m)^\intercal \boldsymbol{x} : \boldsymbol{x} \in \mathcal{E}(\boldsymbol{A}_m; \boldsymbol{a}_m)\} \\
&= n(\boldsymbol{a}_m) - \sqrt{\nabla n(\boldsymbol{a}_m)^\intercal \boldsymbol{A}_m \nabla n(\boldsymbol{a}_m)}.
\end{aligned}$$

This implies by Algorithm 3.12 and $\|\nabla n(\boldsymbol{a}_m)\|_2 = 1$ that

$$\begin{aligned}
(1 - \alpha_m)\varepsilon_m &= \left(1 - \frac{n(\boldsymbol{a}_m) - r}{\sqrt{\nabla n(\boldsymbol{a}_m)^\intercal \boldsymbol{A}_m \nabla n(\boldsymbol{a}_m)}}\right) \frac{\sqrt{\nabla n(\boldsymbol{a}_m)^\intercal \boldsymbol{A}_m \nabla n(\boldsymbol{a}_m)}}{\|\nabla n(\boldsymbol{a}_m)\|_2} \\
&= \sqrt{\nabla n(\boldsymbol{a}_m)^\intercal \boldsymbol{A}_m \nabla n(\boldsymbol{a}_m)} - n(\boldsymbol{a}_m) + r \\
&\geq \delta > 0
\end{aligned}$$

and so the result is proven. □

By Theorem 3.3.1 and Lemma 3.3.3 it follows that there exists some m_0 such that $\omega_m = \min\{\omega_m^g, \omega_m^f\}$ for $m \geq m_0$.

Hence, if an instance of (U) is being solved Algorithm 3.11 reduces to Algorithm 3.13 and ω_m reduces for $m \geq m_0$ to ω_m^f, since there is no ω_m^g in this case. Therefore, by Theorem 3.3.1 it follows that $\omega_m^f \downarrow 0$. Recall now that Algorithm 3.13 applies objective cuts only when $\boldsymbol{a}_m \in \mathcal{L}_n^{\leq}(r)$ and since this is a compact set (it is the first ellipsoid) and f is convex there is a constant L_f such that f is Lipschitz continuous over this set. Hence $\|\boldsymbol{a}_m\|_2 \leq L_f$ and this implies that the upper bound in (3.15) is below $L_f \omega_m^f$ for m sufficiently large.

Considering now the constrained case, i.e. an instance of (C), it follows that if an optimal solution $\boldsymbol{x}_*$ is interior to the feasible set then a similar argument as used in

Lemma 3.3.3 leads to the conclusion that ω_m reduces after finitely many iterations to ω_m^f and the previous reasoning applies. Since this "fundamentally" unconstrained case is easy we turn now our attention to the case where all the optimal solutions satisfy $g(\boldsymbol{x}_*) = 0$.

It turns out that the satisfaction of stopping rules for this case is much more difficult to analyze, despite of the finite satisfaction of the criteria being guaranteed by Theorem 3.2.4.

However, Corollary 3.2.1 can be improved in order to show that under Assumption 3.2.1 an infinite sequence of objective cuts is always generated. Observe, from the example discussed in Figure 3.2, that this result is an effect of the different nature of ellipsoids when compared with intervals.

Before presenting this result we need to extend the analysis of the ellipsoid method. Recall by (3.6) that the center of $\mathcal{E}(\boldsymbol{A}_{m+1}; \boldsymbol{a}_{m+1})$ lies in $[\boldsymbol{a}_m, \boldsymbol{a}_m - \boldsymbol{b}_m]$. It has been already implied by the discussion after (3.7) that $\boldsymbol{a}_m - \boldsymbol{b}_m \in \text{bd}(\mathcal{E}(\boldsymbol{A}_m; \boldsymbol{a}_m))$. This is very easy to prove, by computing $((\boldsymbol{a}_m - \boldsymbol{b}_m) - \boldsymbol{a}_m)^\intercal \boldsymbol{A}^{-1} ((\boldsymbol{a}_m - \boldsymbol{b}_m) - \boldsymbol{a}_m)$ and observing that it equals 1. The next result follows now from each generated ellipsoid having full dimension (although they may converge to a degenerate ellipsoid) and from Lemma 2.3.1, since $\mathcal{E}(\boldsymbol{A}; \boldsymbol{a}) = \mathcal{L}_\epsilon^{\leq}(0)$ with $\epsilon(\boldsymbol{x}) := (\boldsymbol{x} - \boldsymbol{a})^\intercal \boldsymbol{A}^{-1}(\boldsymbol{x} - \boldsymbol{a}) - 1$ a (quadratic) convex function for which $\epsilon(\boldsymbol{a}) = -1$. Hence, by Lemma 2.2.4 and the ellipsoid being full dimensional it follows that $\boldsymbol{a}_m - \mu \boldsymbol{b}_m \in \text{int}(\mathcal{E}(\boldsymbol{A}_m; \boldsymbol{a}_m))$ for every $0 \leq \mu < 1$.

Lemma 3.3.4. *At every iteration m of an ellipsoid algorithm the point $\boldsymbol{z}_m := \boldsymbol{a}_m - \alpha_m \boldsymbol{b}_m$ belongs to $\mathcal{L}_{\bar{h}_m}^{=}(\beta_m)$ and to* $\text{int}(\mathcal{E}(\boldsymbol{A}_{m+1}; \boldsymbol{a}_{m+1}))$.

Proof: The first part, i.e. $\boldsymbol{a}_m - \alpha_m \boldsymbol{b}_m \in \mathcal{L}_{\bar{h}_m}^{=}(\beta_m)$, follows directly by the definitions of $\boldsymbol{b}_m$, α_m and β_m. Consider now the second part. We already observed that $\boldsymbol{z}_m \in \text{int}(\mathcal{E}(\boldsymbol{A}_m; \boldsymbol{a}_m))$ and hence by the first part there exists some $\delta > 0$ such that $(\boldsymbol{z}_m + \delta \mathcal{B}) \cap \mathcal{L}_{\bar{h}_m}^{=}(\beta_m) \subseteq \text{int}(\mathcal{E}(\boldsymbol{A}_{m+1}; \boldsymbol{a}_{m+1}))$. In order to lift the dimension of the $s-1$-dimensional ball $(\boldsymbol{z}_m + \delta \mathcal{B}) \cap \mathcal{L}_{\bar{h}_m}^{=}(\beta_m)$ consider the affine hull of $[\boldsymbol{a}_m, \boldsymbol{a}_m - \boldsymbol{b}_m]$, i.e. the line orthogonal to $\mathcal{L}_{\bar{h}_m}^{=}(\beta_m)$ which intersects it in $\boldsymbol{z}_m$. Observe now that the point $\boldsymbol{a}_{m+1} := \boldsymbol{a}_m - \tau_m \boldsymbol{b}_m$ belongs to this line and also to the interior of $\mathcal{E}(\boldsymbol{A}_{m+1}; \boldsymbol{a}_{m+1})$ since it is the center of this full dimensional and symmetric (around

$\boldsymbol{a}_{m+1}$) convex set. Also $\boldsymbol{a}_m - \boldsymbol{b}_m \in \text{bd}(\mathcal{E}(\boldsymbol{A}_m; \boldsymbol{a}_m)) \cap \mathcal{L}^{\leq}_{\bar{h}_m}(\beta_m) \subseteq \mathcal{E}(\boldsymbol{A}_{m+1}; \boldsymbol{a}_{m+1})$ and from the symmetry of $\mathcal{E}(\boldsymbol{A}_{m+1}; \boldsymbol{a}_{m+1})$ around $\boldsymbol{a}_{m+1}$ it follows that $]\boldsymbol{a}_m - (2\tau_m - 1)\boldsymbol{b}_m, \boldsymbol{a}_m - \boldsymbol{b}_m[\subseteq \text{int}(\mathcal{E}(\boldsymbol{A}_{m+1}; \boldsymbol{a}_{m+1}))$. Now, since $\boldsymbol{z}_m \in]\boldsymbol{a}_m - (2\tau_m - 1)\boldsymbol{b}_m, \boldsymbol{a}_m - \boldsymbol{b}_m[$ and $\boldsymbol{z}_m \in (\boldsymbol{z}_m + \delta\mathcal{B}) \cap \mathcal{L}^{=}_{\bar{h}_m}(\beta_m)$ it follows that

$$\begin{aligned} \boldsymbol{z}_m &\in \text{int}\left(\text{conv}\left(]\boldsymbol{a}_m - (2\tau_m - 1)\boldsymbol{b}_m, \boldsymbol{a}_m - \boldsymbol{b}_m[\cup \left((\boldsymbol{z}_m + \delta\mathcal{B}) \cap \mathcal{L}^{=}_{\bar{h}_m}(\beta_m)\right)\right)\right) \\ &\subseteq \text{int}(\mathcal{E}(\boldsymbol{A}_{m+1}; \boldsymbol{a}_{m+1})) \end{aligned}$$

from where the result follows. □

We need to turn our attention now to the sequence $\boldsymbol{z}_m$. In case $\alpha_m = 0$ this is $\boldsymbol{a}_m$. In the case of the bisection algorithm, one dimension ellipsoid, this point is the new end point of the new localization interval and hence a boundary point. Clearly Lemma 3.3.4 shows the difference it makes to use "round enclosures" as ellipsoids. This is just the observation that leads to the following theorem.

Theorem 3.3.2. *If Assumption 3.2.1 holds the execution of Algorithm 3.11 generates an infinite sequence of objective cuts.*

Proof: The proof follows a similar reasoning as the proof of Corollary 3.2.1 by making use of Lemma 3.3.4. The proof of Corollary 3.2.1 works by putting balls inside the set $\mathcal{L}^{\leq}_{\bar{f}}(\ell_m) \cap \mathcal{L}^{\leq}_{\bar{g}}(0) \cap \mathcal{L}^{\leq}_{\bar{n}}(r)$. This works until $\ell_m = f(\boldsymbol{x}_*)$. Now we prove that the set $\mathcal{L}^{\leq}_{\bar{g}}(0) \cap \mathcal{L}^{\leq}_{\bar{n}}(r) \cap \mathcal{E}(\boldsymbol{A}_k; \boldsymbol{a}_k)$, with k the last iteration until iteration m where an objective cut is performed, is always full dimensional, hence allows a ball inside, and so the same volume argument as in the proof of Corollary 3.2.1 leads to a point inside this set having to be found in a subsequent iteration. Such point originates an objective cut. The only cut which changes $\mathcal{L}^{\leq}_{\bar{g}}(0) \cap \mathcal{L}^{\leq}_{\bar{n}}(r) \cap \mathcal{E}(\boldsymbol{A}_k; \boldsymbol{a}_k)$ is an objective cut. Suppose now that such a cut is performed in iteration $m > k$. If point $\boldsymbol{z}_m \in \text{int}\left(\mathcal{L}^{\leq}_{\bar{g}}(0) \cap \mathcal{L}^{\leq}_{\bar{n}}(r)\right)$ then by Lemma 3.3.4 there exists a $\delta_m > 0$ such that $\boldsymbol{z}_m + \delta_m\mathcal{B} \subseteq \mathcal{L}^{\leq}_{\bar{g}}(0) \cap \mathcal{L}^{\leq}_{\bar{n}}(r) \cap \mathcal{E}(\boldsymbol{A}_{m+1}; \boldsymbol{a}_{m+1})$ and since this set remains unchanged until the next objective cut the result follows. Consider now the case $\boldsymbol{z}_m \in \text{bd}\left(\mathcal{L}^{\leq}_{\bar{g}}(0) \cap \mathcal{L}^{\leq}_{\bar{n}}(r)\right)$. By a similar argument as used in the discussion after Assumption 3.2.1 it follows that a point $\tilde{\boldsymbol{z}}_m$ exists in int $\left(\mathcal{L}^{\leq}_{\bar{g}}(0) \cap \mathcal{L}^{\leq}_{\bar{n}}(r)\right)$ together with a $\tilde{\delta}_m > 0$ such that $\tilde{\boldsymbol{z}}_m + \tilde{\delta}_m\mathcal{B} \subseteq \boldsymbol{z}_m + \delta_m\mathcal{B} \cap \text{int}(\mathcal{L}^{\leq}_{\bar{g}}(0) \cap \mathcal{L}^{\leq}_{\bar{n}}(r)) \cap \text{int}(\mathcal{E}(\boldsymbol{A}_{m+1}; \boldsymbol{a}_{m+1}))$ and this proves the result. □

Hence, also in the constrained case, the stopping rule is evaluated (in the worst case) infinitely many times. However, we cannot prove that ω_m^f converges to zero.

The next section describes the way our computational experiments were conducted. The results of these experiments are reported in Chapter 4 and in Chapter 6.

3.4. Computational experience

Ecker and Kupferschmid tested the ellipsoid method (Algorithm 3.6) in Ref. [31] against several well-established optimization methods in differentiable nonlinear programming. Their computational study was carried out over convex and nonconvex problems with several local minima. According to Ecker and Kupferschmid the ellipsoid method is a competitive, stable, efficient and multipurpose method in nonlinear programming. Moreover, they observe that in several cases where no optimal solution is contained in the initial ellipsoid the method could often find one. This is a side effect of the fact that each new ellipsoid is not completely included in the previous one. Observe that the same cannot happen with Algorithm 3.11 since the use of norm cuts prohibits the consideration of centers outside the initial ellipsoid. Ecker and Kupferschmid also remark that for some well-known test functions possessing several local minima the ellipsoid method exhibits a notorious trend to pick the lowest of them.

In our experience we rule out of consideration cases of "good luck" such as finding points outside the first ellipsoid or the best of several local minima by focusing only on instances which satisfy Assumption 3.1.1 and have convex lower level sets.

These instances are provided by convex location models presented and discussed in Chapter 4 and by quasiconvex location models presented and discussed in Chapter 6.

We now describe how the computational experiments whose results are reported in this book was carried out. In order to test the different algorithms they were completely coded by us in Turbo Pascal version 7.0 and no commercial routines were used except the standard functions and procedures of the language. The programs were executed on a Compaq Prolinea 4/66, a PC/AT compatible with an Intel 80486 DX2 processor with built in numerical processor and clock speed of 66 MHz. The numerical precision used in most of the cases is the *Double Precision* (64-bit IEEE floating point format) real numbers of Turbo Pascal. The only exception makes use of the Turbo Pascal specific *Extended Precision*, a non IEEE 80-bit numerical format superior in precision to the IEEE 64-bit Double Precision format. This exception corresponds to the tests reported in Subsection 4.6.3.

As a general policy adopted in our experiments for each problem 10 similar but uncorrelated instances were generated, and hence rows in tables correspond to averages of values obtained for each of those 10 instances.

Details on how each instance is generated are listed in each section where computational results are presented.

IV Convexity in Location

We start this chapter by introducing in Section 4.1 some well-known notions in single facility continuous location. Since distances play a prominent role in this field of nonlinear programming, Section 4.2 elaborates on ways to measure distance, enlightening the role of gauges. Also in this section a geometrical interpretation of the meaning of some important gauges is presented. A mathematical description of a general single facility unconstrained continuous location model is then considered in Section 4.3. It is also shown that the proposed model fulfills Assumption 3.1.1, which is essential to enable the application of the ellipsoid algorithm. In Section 4.4 the separation oracle required to execute the ellipsoid algorithm described in the previous chapter is detailed for a restricted subset of the above model, while in Section 4.5 the model is further restricted so that the identification of optimal solutions can be stated as workable instances of a so-called *membership problem*. The planar case of this membership problem is analyzed in Section 4.6. In this section algorithms are derived (in Subsection 4.6.1 and Subsection 4.6.2) to solve two instances of this decision problem when ℓ_p-norms and/or polyhedral gauges are used to measure distance. Finally, some computational results are presented.

4.1. Introduction

Location problems concern the determination of *location(s)* or *site(s)* for one or more so-called *facilities* taking into account a given set of points, the so-called *demand points*, with which *interactions* are to be established. These terms, not being part of a standard terminology, are sometimes replaced by others. Typical alternatives for the term demand points are *clients*, *existing facilities*, *users* or *markets*. These terms may induce the idea that the service, whatever that may be, provided by the facility or facilities is always a desirable one, but that is not the case. It depends on the nature of the interactions between facilities and demand points, which are represented in some applications by transportation of goods, services offered or requested, physical links to be explored or constructed, or undesirable influences such as contamination or pollution. Some military applications also consider as interactions the risk of detection of secret material. For references to models exploring all these aspects of location problems we refer to the extensive survey by Plastria presented in Ref. [82].

Location problems are termed *continuous* when the underlying space for both facilities and demand points is a continuous one. More specifically, all points defining an instance of the problem are given by one or more coordinates (depending on the dimension) which may vary continuously.

This book does not deal with particular models or interpretations. Moreover, we concentrate on continuous instances occurring in the Euclidean space $\mathbb{R}^s$ of dimension at least $s = 2$. Dimension $s = 2$, the so-called *planar case*, is the adequate setting for the vast majority of situations for obvious reasons of geographical nature. We also concentrate on locating a *single facility*[1]. Therefore, we adopt from now on exclusively the terminology *facility* and *demand points.*

The interactions to be taken into account are also understood in this book as being a function of the spatial position of the involved points measured by means of some *distance.* A global assumption we make about such distance measures is convexity. We will clarify this point in the next section.

In order to define an instance of a continuous location problem we first need to introduce a set $\mathcal{D} := \{\boldsymbol{d}_1, \ldots, \boldsymbol{d}_n\} \subseteq \mathbb{R}^s$ of n distinct demand points.

An example for $s = 2$, the so-called *planar* case, is provided by the first picture in Figure 4.1 where the 10 small squares represent the 10 demand points in this instance.

Given these demand points one needs to find an optimal location $\boldsymbol{x} \in \mathbb{R}^s$ for a facility, such that this location minimizes some real-valued function involving the distances between this new location and each one of the demand points.

[1]Notice that a common technique for accommodating more facilities consists on increasing the dimension of the problem in such a way that the single "facility" to locate is nothing else than the list of coordinates of all the facilities of the underlying model.

An example of a solution is given by the second picture of Figure 4.1 where the facility is represented by the small circle and the lines represent the distances between each pair (facility,demand point).

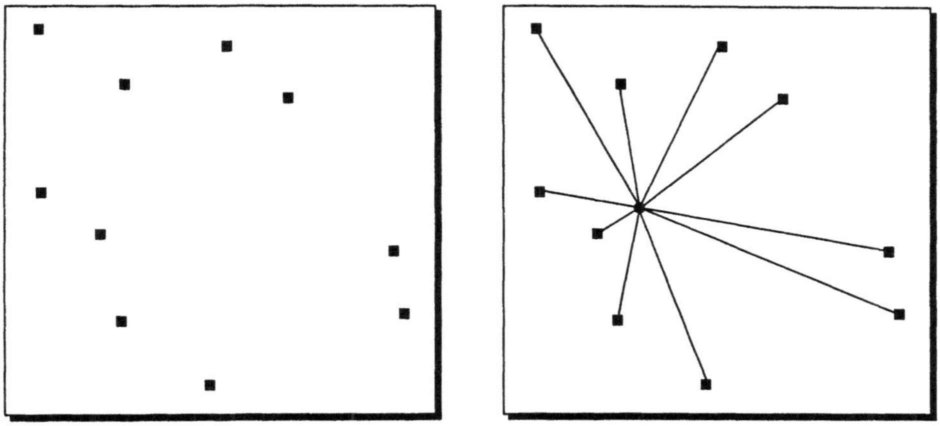

Figure 4.1: The single facility planar location problem

These distances are measured in very general ways. In the next section we present suitable techniques to measure the so-called *convex distances.*

4.2. Measuring convex distances

A distance measure on $\mathbf{R}^s$ is generally understood as a function $d : \mathbf{R}^s \times \mathbf{R}^s \longrightarrow \mathbf{R}_+$ satisfying a group of properties. Mostly, these properties are those required for a *metric.* We prefer not to assume symmetry but in order to enable the use of convex analysis (the main analytical tool in nonlinear programming) we demand $d(x, \cdot)$ and $d(\cdot, y)$ to be convex functions. This, in combination with the assumptions made on a metric without symmetry, was proven by Witzgall (Ref. [111]) to imply that d is derived from a *gauge.* Let $\mathcal{G} \subseteq \mathbf{R}^s$ be a compact convex set with $\mathbf{0} \in \mathrm{int}(\mathcal{G})$. A *gauge* or *Minkowsky functional* is defined in Chapter 2 for each $x \in \mathbf{R}^s$ as

$$\gamma_{\mathcal{G}}(x) := \inf\{t > 0 : x \in t\mathcal{G}\}.$$

Moreover, if $\mathcal{G}$ is symmetric around $\mathbf{0}$ the corresponding gauge is called a *norm.*

In Chapter 2 we elaborate on geometrical properties of epigraphs of gauges. This section is more concerned with properties of gauges at the "domain level". An immediate geometric observation is that all the lower level sets of $\gamma_{\mathcal{G}}$ "look like" $\mathcal{G}$. In fact, they are enlarged or reduced replicas of $\mathcal{G}$. In this way one can view a gauge as the function that returns the value by which $\mathcal{G}$ should be multiplied (if this value is smaller than 1 the set reduces, if it is greater than 1 the set enlarges) in such a way that the argument point belongs to the boundary of the resulting set.

We now proceed with some examples.

The most well-know norm is without doubt the so-called *Euclidean norm* which is defined by

$$\|\boldsymbol{x}\|_2 = \left(\sum_{j=1}^{s} x_j^2\right)^{1/2} = \sqrt{\boldsymbol{x}^\top \boldsymbol{x}}.$$

This norm corresponds to the gauge of the set $\mathcal{G}_2 := \{\boldsymbol{x} \in \mathbf{R}^s : \|\boldsymbol{x}\|_2 \leq 1\}$. This is a sphere (or a circle if $s = 2$) centered in $\mathbf{0}$ and with radius equal to 1. An illustration for the planar case, $s = 2$, is provided by the first picture in Figure 4.2. By analogy with this norm the set $\mathcal{G}$ of a gauge $\gamma_{\mathcal{G}}$ is always called the *unit circle* of this gauge.

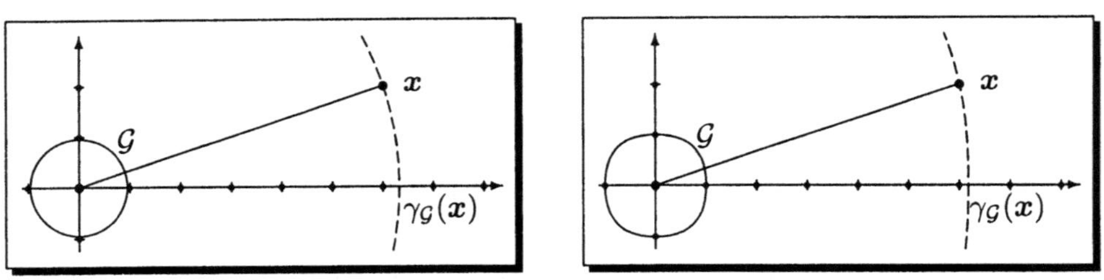

Figure 4.2: The Euclidean norm and the $\ell_{2.4}$-norm

Other norms related to the Euclidean (also called ℓ_2) norm are the ℓ_p*-norms*. These norms are defined for $1 < p < \infty$ by

$$\|\boldsymbol{x}\|_p := \left(\sum_{j=1}^{s} |x_j|^p\right)^{1/p}.$$

Similarly to the Euclidean norm these norms are gauges of the sets $\mathcal{G}_p := \{x \in \mathbb{R}^s : \|x\|_p \leq 1\}$. An example for the planar case with $p = 2.4$ is provided by the second picture in Figure 4.2.

The cases $p = 1$ and $p = \infty$ are limit cases of ℓ_p-norms and correspond to

$$\|x\|_1 := \sum_{j=1}^{s} |x_j| \qquad \text{and} \qquad \|x\|_\infty := \max\{|x_j| : 1 \leq j \leq s\}.$$

Observe that these two norms are generally included in the ℓ_p family but we prefer to isolate them due to fact that their gauges are defined by sets of a different nature than the ones defining ℓ_p-norms for $1 < p < \infty$. These two norms are also known as *polyhedral gauges*, i.e. gauges for which $\mathcal{G}$ is a polytope. Polyhedral gauges for symmetric polytopes are also known as *block norms* (Ref. [108, 109]).

Two examples of these gauges are presented in Figure 4.3.

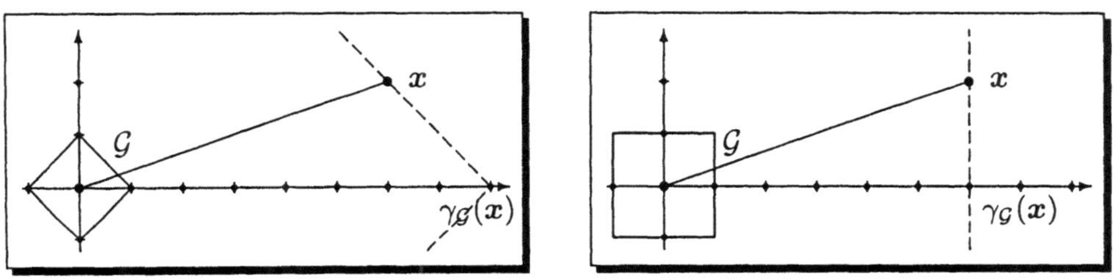

Figure 4.3: Two polyhedral gauges: Rectangular (ℓ_1) and Tchebycheff (ℓ_∞) norms

Polyhedral gauges have an interesting interpretation when used to measure distances (Ref. [109]). This interpretation also holds for the asymmetric case. An example of such a gauge is provided in Figure 4.4.

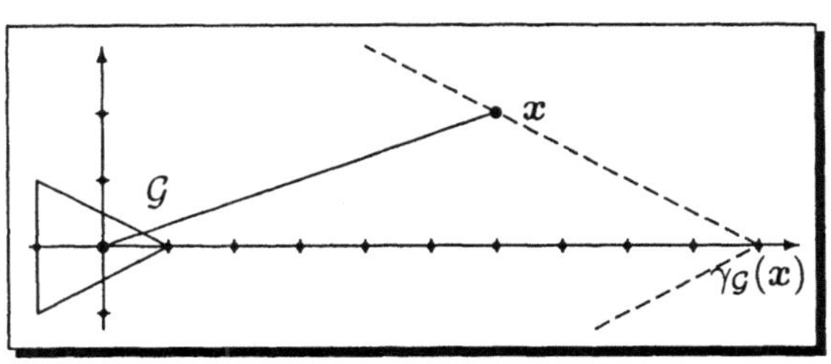

Figure 4.4: An asymmetrical gauge

The value of a polyhedral gauge corresponds to the *travel time* along the *fastest* path in a dense network of one-ways. Each way is oriented by a position vector of a vertex of $\mathcal{G}$ and the travel speed along it is proportional to the Euclidean length of this vector.

Let us assume in Figure 4.4 that the Euclidean length of the position vector of the rightmost vertex is 1 and exhibit the underlying network in Figure 4.5. From the Theorem of Pythagoras we obtain that the length of each of the other two position vectors is given by $\sqrt{1^2+1^2}$. Hence, this gauge represents a grid of three different paths, one traveling from left to right at speed 1, the other from down right to up left at speed $\sqrt{2}$ and the third one from up right to down left also at speed $\sqrt{2}$.

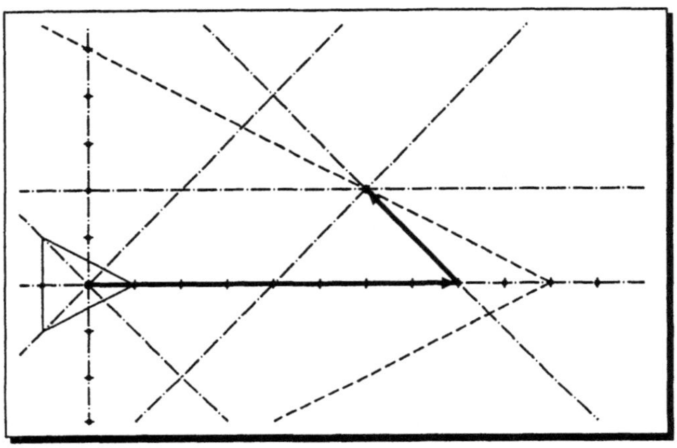

Figure 4.5: Interpretation of a polyhedral gauge

Now, in order to travel from $\mathbf{0}$ to x there are many feasible ways. One such way (which corresponds to one of the possibilities to realize the fastest trip from $\mathbf{0}$ to x) is represented in Figure 4.5 and corresponds to first traveling at speed 1 along the line generated by the position vector of the rightmost vector for 8 length units and so taking 8 units of time. From this point on the line generated by the position vector of the top left vertex is followed at speed equal to $\sqrt{2}$ and (applying again the Theorem of Pythagoras) for a length of $\sqrt{2^2+2^2}$.

Thus, the total time required to travel from $\mathbf{0}$ to x equals

$$8+\frac{\sqrt{2^2+2^2}}{\sqrt{1^2+1^2}}=8+\sqrt{\frac{8}{2}}=8+2$$

and this is the value of the gauge at x.

Observe that traveling from x to $\mathbf{0}$ takes 6 units of time instead of 10 and this value equals $\gamma_{\mathcal{G}}(-x)$. The equality $\gamma_{\mathcal{G}}(x) = \gamma_{\mathcal{G}}(-x)$ only holds for $\mathcal{G}$ symmetric.

So far we have seen that a gauge $\gamma_{\mathcal{G}}$, when evaluated in x, measures the "distance" from $\mathbf{0}$ to x. In order to measure the distance from a point x to a point y we simply need to compute $\gamma_{\mathcal{G}}(y - x)$ using an appropriate gauge.

Observe that Figure 4.2, Figure 4.3 and Figure 4.4 are drawn with equal scale and that in all cases the same point x is used. However, the value of the gauge at x is different for each case.

The motivation for considering continuous location problems with distances derived from arbitrary gauges is twofold. First of all, although most papers only discuss continuous location models with Rectangular, Euclidean or Tchebycheff norms, it is known that an ℓ_p-norm with p not equal to 1, 2 or ∞ or a general polyhedral or block norm (Ref. [55, 109]) sometimes fits better to empirical data (Ref. [13, 66, 108, 109]). This implies that it may be sensible to consider arbitrary norms within location models. Secondly, it is sometimes unrealistic to assume, as imposed by a norm, that the distance is symmetric (Ref. [82]). Hence, to model this asymmetric behavior of distances, i.e. the distance between x and y is not equal to the distance between y and x, we need to consider arbitrary gauges. It should be clear from the above observations that the selection of a distance function is central in all continuous location problems and this topic is intensively discussed in the recent survey paper by Plastria (Ref. [83]). To conclude our discussion on how to measure distance we observe that the Euclidean, Rectangular or Tchebycheff norms were mainly used to obtain easy tractable nonlinear/linear programming problems (Ref. [67]).

The importance of polyhedral gauges is also twofold. First, since polytopes form a dense subset of the set of compact convex sets (each compact convex set can be approximated to any degree by such a polytope) they generate a dense family of gauges. On the other hand, for some practical situations they offer exact measurements of the distance involved.

An example is provided by the field of robotics, see Ref. [33]. Let us restrict to a situation where movement is allowed only in two directions and generated by one motor[2] per direction. Examples of such devices are some harbor cranes used to load and unload containers, robot arms for insertion of components in printed boards and plotters as the one represented in Figure 4.6.

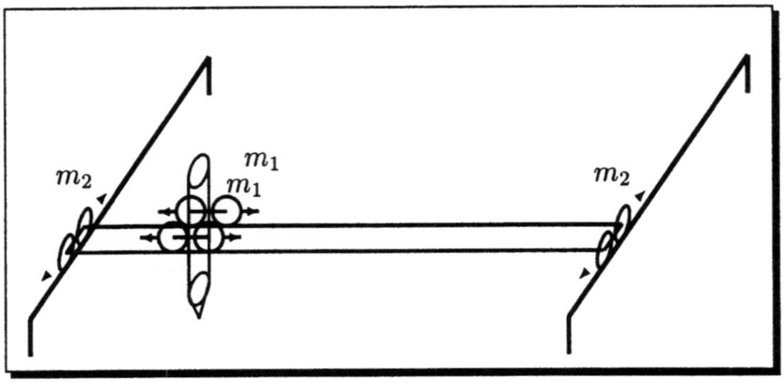

Figure 4.6: An example of robotic distance: a plotter

We assume that the motors move at constant speed and that the speed is the same for both motors. Without loss of generality we consider this speed to be 1. Suppose first that the motors always move simultaneously and at speed 1. In this case the plotter's pen travels along the bisector of each quadrant at speed $\sqrt{2}$ since it moves from $(0,0)$ to $(1,1)$ in time 1. Hence the distance between two points accessible by the pen's point is appropriately measured by the ℓ_∞-norm. On the other hand, if the motors move only sequentially then the adequate norm becomes the ℓ_1-norm.

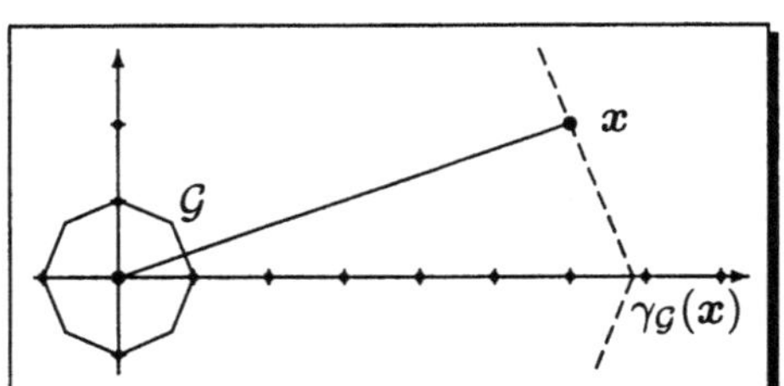

Figure 4.7: An example of a one-infinity norm

Suppose now that a combination of strategies is allowed and that the electric motors are connected in series. In this case if any of the motors is the only one moving it

[2]One "motor" can physically be implemented by several motors, see Figure 4.6 where the motors are marked as m_1 and m_2.

moves at speed 1 since it receives the total current, but if both move the current is divided and the speed of each is halved. It follows that the pen travels along the coordinated directions at speed one and along the bisectors at speed $\sqrt{2}/2$, since in time 1 it moves from $(0,0)$ to $(1/2,1/2)$. This makes one of the so-called *one-infinity norms* (Ref. [108]) the adequate one. An example of such a norm is given in Figure 4.7.

By means of the following figure we illustrate the three different strategies mentioned before for moving the plotter's pen. Suppose the pen is at $\boldsymbol{a}$ and should be moved to $\boldsymbol{b}$. If the motors can only move one at the time, sequentially as described by the ℓ_1-norm, then the pen moves first from $\boldsymbol{a}$ to $\boldsymbol{c}$ and then from $\boldsymbol{c}$ to $\boldsymbol{b}$. On the other hand, if the motors can only be moved simultaneously as described by the ℓ_∞-norm, then the pen moves first from $\boldsymbol{a}$ to $\boldsymbol{d}$ and then from $\boldsymbol{d}$ to $\boldsymbol{b}$. Finally, if the previous strategies may be combined in a one-infinity fashion then the pen moves first from $\boldsymbol{a}$ to $\boldsymbol{e}$ and then from $\boldsymbol{e}$ to $\boldsymbol{b}$. Observe that in order to mimic the behavior of a smooth norm, e.g. the Euclidean norm, the plotter should be able to set the speed of each motor from a continuous range of possibilities.

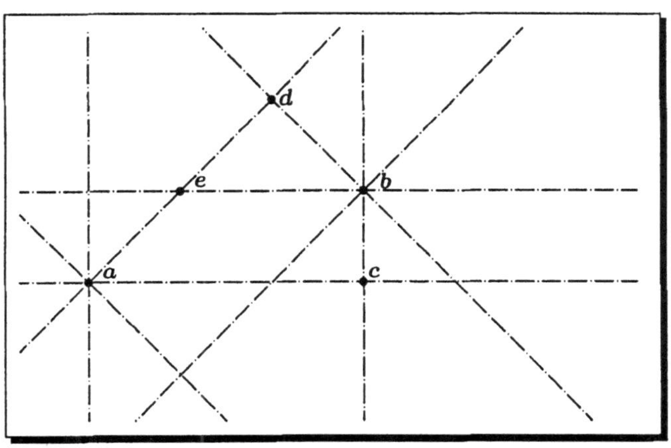

Figure 4.8: Some possible movements of a plotter's pen

It is curious to observe that nothing new is obtained by assuming that the motors are connected in parallel and can move independently. In this case a possible strategy would be moving first from $\boldsymbol{a}$ to $\boldsymbol{e}$ at speed $\sqrt{2}$ and from here to $\boldsymbol{b}$ at speed one, however the time taken to reach $\boldsymbol{b}$ equals the one measured by the ℓ_∞-norm since the same time is taken by moving first from $\boldsymbol{a}$ to $\boldsymbol{d}$ and from here to $\boldsymbol{b}$ always at speed $\sqrt{2}$. This is just as if the unit ℓ_∞-circle would have four additional "vertices" in $\boldsymbol{e}_1$, $\boldsymbol{e}_2$, $-\boldsymbol{e}_1$ and $-\boldsymbol{e}_2$ with $\boldsymbol{e}_i$ the ith unit vector. This interpretation suggests that the unit ℓ_∞-circle is the outer limit of the one-infinity family, see Ref. [108]. Similarly the ℓ_1-circle appears as the inner limit. Figure 4.9 illustrates this property.

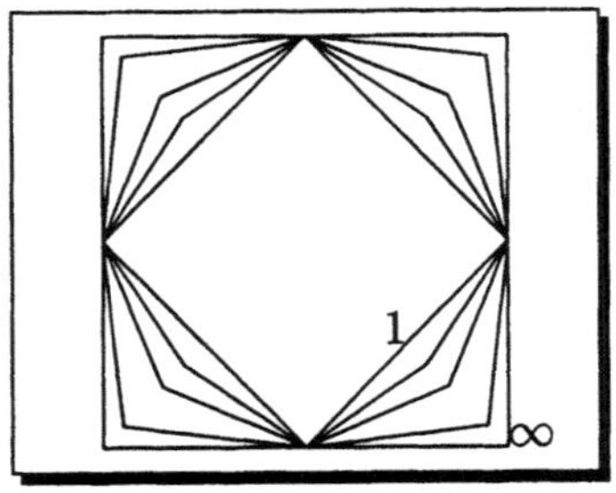

Figure 4.9: The one-infinity family

This concludes our discussion about properties of gauges. In the next section we consider a mathematical description of a general single facility location model.

4.3. A general model

If $\mathcal{G}_i$, $i = 1, \ldots, n$, are compact convex sets with $\mathbf{0} \in \text{int}(\mathcal{G}_i)$ and $\text{int}(\mathcal{G}_i)$ denoting the interior of $\mathcal{G}_i$, we define the distance between $\boldsymbol{d}_i$ and $\boldsymbol{x}$ by $\gamma_{\mathcal{G}_i}(\boldsymbol{x} - \boldsymbol{d}_i)$ with $\gamma_{\mathcal{G}_i}$ the *gauge* or Minkowsky functional of the set $\mathcal{G}_i$, see (2.3) and the previous section.

To introduce the model, let $\gamma : \mathbf{R}^s \longrightarrow \mathbf{R}^n_+$ be a vector function given by

$$\gamma(\boldsymbol{x}) := (\gamma_{\mathcal{G}_1}(\boldsymbol{x} - \boldsymbol{d}_1), \ldots, \gamma_{\mathcal{G}_n}(\boldsymbol{x} - \boldsymbol{d}_n)) \tag{4.1}$$

and suppose $\varphi_i : \mathbf{R}^n_+ \longrightarrow \mathbf{R}_+$, $1 \leq i \leq n$, denote a set of nondecreasing so-called *disutility functions* on $\mathbf{R}^n_+$ associated with the demand points $\boldsymbol{d}_i$. We also assume that the functions φ_i are continuous on an open set $\mathcal{S}$ with $\mathbf{R}^n_+ \subseteq \mathcal{S}$.

The single facility location model discussed in this book is now defined by

$$(L) \quad \inf_{\boldsymbol{x} \in \mathbf{R}^s} \max_{1 \leq i \leq n} f_i(\boldsymbol{x})$$

with $f_i : \mathbf{R}^s \longrightarrow \mathbf{R}$ given by $f_i(x) := \varphi_i(\gamma(x))$. The above optimization problem is called a min-max problem and arises for example in the location of an emergency unit (Ref. [67]). Clearly, if all the functions φ_i are equal, problem (L) reduces to $\inf\{f_1(x) : x \in \mathbf{R}^s\}$. Most continuous unconstrained single facility location problems are special cases of the optimization problem (L) (Ref. [67]) and this problem has in general a nondifferentiable objective function even if the disutility functions φ_i are assumed differentiable for each $i = 1, \ldots, n$ (Ref. [91]). Since the functions φ_i are continuous on the open set $\mathcal{S}$ and nondecreasing on $\mathbf{R}^n_+$ it follows that the function $z \longmapsto \max_{1 \leq i \leq n} \varphi_i(z)$ satisfies the same properties and hence our model is a special case of a more general model discussed in Ref. [79]. This implies (Ref. [79, 83]) that there exists an easily computable hyperrectangle $\mathcal{R}$ which contains at least one optimal solution of (L). Moreover, by Ref. [79] this hyperrectangle equals $\mathcal{R}(l, u) := \{x \in \mathbf{R}^s : l \leq x \leq u\}$ with

$$l_j := \min_{1 \leq i \leq n} \max_{1 \leq k \leq n} \left(d_{k_j} + \gamma_{\mathcal{G}_k}(d_k - d_i)\gamma_{\mathcal{G}_k^\circ}(e_j)\right)$$

and

$$u_j := \max_{1 \leq i \leq n} \min_{1 \leq k \leq n} \left(d_{k_j} - \gamma_{\mathcal{G}_k}(d_k - d_i)\gamma_{\mathcal{G}_k^\circ}(e_j)\right)$$

for each $1 \leq j \leq s$ and $\gamma_{\mathcal{G}_k^\circ}$ denoting the dual gauge of $\gamma_{\mathcal{G}_k}$. If the distance is measured by the same arbitrary norm for each demand point and $s = 2$, this hyperrectangle $\mathcal{R}$ can be replaced by the smaller set conv$(\mathcal{D})$, with conv$(\mathcal{D})$ the convex hull of $\mathcal{D}$ (Ref. [110]) or by any rectangle containing conv$(\mathcal{D})$. From these remarks it now follows that there exists an easily computable hyperrectangle $\mathcal{R}$ such that solving (L) is equivalent to solving the constrained problem

$$(K) \quad \min\{f(x) : x \in \mathcal{R}\}$$

with $f(x) := \max_{1 \leq i \leq n} f_i(x)$. From the equivalence between (K) and (L), it is also easy to find some $a_0 \in \mathbf{R}^s$ and $r > 0$ such that an optimal solution x_* belongs to the set $a_0 + r\mathcal{B}$ with $\mathcal{B}$ the unit Euclidean ball. A way to construct such a sphere consists of taking a_0 as the center of $\mathcal{R}$ and r bigger than the Euclidean distance from a_0 to one of the rectangle's vertices. This means that (K) satisfies Assumption 3.1.1 and this sphere yields a starting ellipsoid for Algorithm 3.13.

However, the model described so far does not offer a way to implement a separation oracle. Observe that without such an oracle the ellipsoid method cannot be applied. The next section restricts the class of functions φ_i in a sufficient way to ensure the availability of such an oracle. In Chapter 6 this class is substantially enlarged.

4.4. A convex location model

Let us additionally assume that $\varphi_i : \mathbf{R}^n \longrightarrow \mathbf{R}$ is a finite-valued convex function for each $i = 1, \ldots, n$.

By Theorem 2.3.10 this assumption suffices to guarantee the convexity of f. As a result, this already provides the theoretical availability of a separation oracle, since the function f is finite valued on $\mathbf{R}^s$ and hence $\partial f(\boldsymbol{x})$ is nonempty for every $\boldsymbol{x}$. In order to implement such a separation oracle we need now a way to extract an element of $\partial f(\boldsymbol{a}_m)$ for each $\boldsymbol{a}_m$, see Algorithm 3.13.

Observe that Theorem 2.3.10 combined with Corollary 2.3.3 in Subsection 2.3.5 gives a way to construct an element of $\partial f(\boldsymbol{a}_m)$.

The following algorithm details the procedure in the framework of Algorithm 3.13.

select i such that $\varphi_i(\gamma(\boldsymbol{a}_m)) = f(\boldsymbol{a}_m)$;
take $\boldsymbol{\xi}^\star \in \partial\varphi_i(\gamma(\boldsymbol{a}_m))$;
for $j := 1$ *to* n *do* take $\boldsymbol{\nu}_j^\star \in \partial\gamma_{G_j}(\boldsymbol{a}_m - \boldsymbol{d}_j)$;
let $\boldsymbol{a}_m^\star := \sum_{j=1}^n \xi_j^\star \boldsymbol{\nu}_j^\star$

Algorithm 4.1: Separation oracle for the convex location problem

As this oracle shows, the reasoning repeats itself and is built up by smaller separation oracles from the convex components of the function.

The first framed statement is clearly function dependent. For the sake of this section it is enough to assume here that the disutility functions belong to some "nice" family of convex functions for which such an oracle is easily at hand.

Regarding the second framed statement, which is executed once for each demand point and corresponding gauge, we refer to the discussion on gauges in Chapter 2.

4.5. Characterizing optimality

In order to decide upon optimality of the present iterate of Algorithm 3.13 we need to characterize $\partial f(a_m)$ entirely. This is undoubtedly too difficult for a general family of convex disutility functions φ_i, as Theorem 2.3.10 shows.

Assuming φ_i to be differentiable for each $i = 1, \ldots, n$, as in Ref. [40], is obviously a very useful and appealing simplification but too restrictive from a modeling point of view, since it even excludes very simple functions as the piecewise linear ones. However, from Theorem 2.3.4 it turns out that a convex function fails to be differentiable only on a zero (Lebesgue) measure subset of the interior of its domain. Hence, complementing the traditional view of a convex function as the supremum of possibly infinitely many affine functions, it may also be viewed as the maximum of differentiable convex functions. Here, we simplify by letting each φ_i be given as the maximum of a finite number of known differentiable convex functions, say

$$\varphi_i(z) := \max_{1 \le j \le \kappa_i} \varphi_{i_j}(z).$$

This significantly enlarges the modeling freedom yielding instances of (L). The remainder of this section shows that the analysis of this model is quite simple.

We start by giving a more appealing form to the model by exploiting the associative property of the finite maximum operator.

Let us take $\kappa := \sum_{i=1}^{n} \kappa_i$ and reorder the indices of φ_{i_j} in such a way that

$$\varphi(z) = \max_{1 \le i \le n} \max_{1 \le j \le \kappa_i} \varphi_{i_j}(z) = \max_{1 \le k \le \kappa} \varphi_k(z)$$

The following theorem is now an immediate consequence of Theorem 2.3.7, Theorem 2.3.10, Corollary 2.3.3 and the differentiability of each φ_k with $k = 1, \ldots, \kappa$.

Theorem 4.5.1. *If $\mathcal{C}(\boldsymbol{x}) \subseteq \mathbf{R}^s$ denotes the set*

$$\operatorname{conv}\left(\bigcup_{k\in K(\boldsymbol{x})}\left(\sum_{j=1}^{n}\tfrac{\partial\varphi_k}{\partial z_j}(\gamma(\boldsymbol{x}))\partial\gamma_{\mathcal{G}_j}(\boldsymbol{x}-\boldsymbol{d}_j)\right)\right)$$

with $K(\boldsymbol{x}) := \{k : \varphi_k(\gamma(\boldsymbol{x})) = \varphi(\gamma(\boldsymbol{x})),\ k = 1,\ldots,\kappa\}$ the so-called active set then $\boldsymbol{x}_0$ minimizes the objective function f if and only if $\mathbf{0} \in \mathcal{C}(\boldsymbol{x}_0)$.

To check the above condition for some $\boldsymbol{x}_0$ we need to solve the membership problem

$$\mathbf{0} \overset{?}{\in} \operatorname{conv}\left(\bigcup_{k\in K(\boldsymbol{x}_0)}\left(\sum_{j=1}^{n}\tfrac{\partial\varphi_k}{\partial z_j}(\gamma(\boldsymbol{x}_0))\partial\gamma_{\mathcal{G}_j}(\boldsymbol{x}_0-\boldsymbol{d}_j)\right)\right).$$

In the next examples this membership problem is specified for gauges often used in location theory.

Example 4.5.1. *Taking for each $1 \leq j \leq n$ the set $\mathcal{G}_{p_j}$ defined by $\mathcal{G}_{p_j} := \{\boldsymbol{x} \in \mathbf{R}^s : n_{p_j}(\boldsymbol{x}) \leq 1\}$, $1 < p_j < \infty$, with $n_p(\boldsymbol{x}) := \|\boldsymbol{x}\|_p$ the ℓ_p-norm at $\boldsymbol{x}$, it follows by the above result for $\boldsymbol{x}_0 \notin \mathcal{D}$ that $\boldsymbol{x}_0$ minimizes the objective function f if and only if $\mathbf{0} \in \operatorname{conv}(\{\boldsymbol{c}_k : k \in K(\boldsymbol{x}_0)\})$ with*

$$\boldsymbol{c}_k := \sum_{j=1}^{n}\tfrac{\partial\varphi_k}{\partial z_j}(\gamma(\boldsymbol{x}_0))\nabla n_{p_j}(\boldsymbol{x}_0-\boldsymbol{d}_j)$$

and $\nabla n_{p_j}(\boldsymbol{x})$ the gradient of n_{p_j} at $\boldsymbol{x} \neq \mathbf{0}$. It is well-known for $\boldsymbol{x} \neq \mathbf{0}$ that

$$\nabla n_p(\boldsymbol{x}) = \|\boldsymbol{x}\|_p^{1-p}\begin{bmatrix}\operatorname{sign}(x_1)|x_1|^{p-1}\\ \vdots \\ \operatorname{sign}(x_s)|x_s|^{p-1}\end{bmatrix}$$

with $\operatorname{sign}(x)$ the sign function introduced in Chapter 2. In this case one has to decide upon the feasibility of the linear system

$$\left\{\sum_{k\in K(\boldsymbol{x}_0)}\lambda_k\boldsymbol{c}_k = \mathbf{0},\ \sum_{k\in K(\boldsymbol{x}_0)}\lambda_k = 1,\ \lambda_k \geq 0 \text{ for every } k \in K(\boldsymbol{x}_0)\right\}.$$

A possible way to do this is by applying phase one of the simplex algorithm. Moreover, for $\boldsymbol{x}_0$ belonging to $\mathcal{D}$, say $\boldsymbol{x}_0 = \boldsymbol{d}_\ell$, we have that $\boldsymbol{x}_0$ minimizes the objective function f if and only if $\mathbf{0} \in \text{conv}\left(\bigcup_{k\in K(\boldsymbol{x}_0)} \mathcal{C}_k\right)$ with

$$
\begin{aligned}
\mathcal{C}_k &:= \sum_{j=1}^{n} \tfrac{\partial \varphi_k}{\partial z_j}(\gamma(\boldsymbol{x}_0))\partial\gamma_{\mathcal{G}_j}(\boldsymbol{x}_0 - \boldsymbol{d}_j) \\
&= \sum_{j\neq \ell} \tfrac{\partial \varphi_k}{\partial z_j}(\gamma(\boldsymbol{x}_0))\nabla n_{p_j}(\boldsymbol{x}_0 - \boldsymbol{d}_j) + \tfrac{\partial \varphi_k}{\partial z_\ell}(\gamma(\boldsymbol{x}_0))\partial\gamma_{\mathcal{G}_\ell}(\mathbf{0}) \\
&= \left\{\boldsymbol{x} \in \mathbf{R}^s : \left\|\boldsymbol{x} - \textstyle\sum_{j\neq\ell} \tfrac{\partial \varphi_k}{\partial z_j}(\gamma(\boldsymbol{x}_0))\nabla n_{p_j}(\boldsymbol{x}_0 - \boldsymbol{d}_j)\right\|_{q_\ell} \leq \tfrac{\partial \varphi_k}{\partial z_\ell}(\gamma(\boldsymbol{x}_0))\right\}
\end{aligned}
$$

and $\frac{1}{q_\ell} + \frac{1}{p_\ell} = 1$. *To see this recall Example 2.3.1 and observe that, if $\mathcal{G}_q$ denotes the unit ℓ_q-circle, then $\mathbf{c} + r\mathcal{G}_q = \{\boldsymbol{x} \in \mathbf{R}^s : \|\boldsymbol{x} - \mathbf{c}\|_q \leq r\}$. If $K(\boldsymbol{x}_0)$ contains only one element the above membership problem becomes trivial to check. To conclude this example we observe for $s = 2$ that efficient algorithms are presented in Section 4.6 to solve both membership problems.*

Example 4.5.2. *Consider for each $1 \leq j \leq n$ a polytope $\mathcal{G}_j \subseteq \mathbf{R}^s$, i.e.*

$$\mathcal{G}_j = \text{conv}\left(\{\boldsymbol{v}_{1_j}, \ldots, \boldsymbol{v}_{r_j}\}\right)$$

with $\boldsymbol{v}_{1_j}, \ldots, \boldsymbol{v}_{r_j}$ the vertices of $\mathcal{G}$ and $\mathbf{0} \in \text{int}(\mathcal{G}_j)$. If $\boldsymbol{v}^\circ_{1_j}, \ldots, \boldsymbol{v}^\circ_{r^\circ_j}$ are the vertices of its polar $\mathcal{G}^\circ_j$ it follows again by Theorem 4.5.1 and using Example 2.3.2 that $\boldsymbol{x}_0$ minimizes the objective function f if and only if $\mathbf{0} \in \text{conv}\left(\bigcup_{k\in K(\boldsymbol{x}_0)} \mathcal{C}_k\right)$ with

$$\mathcal{C}_k := \sum_{j=1}^{n} \tfrac{\partial \varphi_k}{\partial z_j}(\gamma(\boldsymbol{x}_0))\, \text{conv}\left(\{\boldsymbol{v}^\circ_\ell : \ell \in I_j(\boldsymbol{x}_0)\}\right),$$

and $I_j(\boldsymbol{x}_0) := \{1_j \leq i \leq r^\circ_j : \gamma_{\mathcal{G}_j}(\boldsymbol{x}_0 - \boldsymbol{d}_j) = \boldsymbol{v}^{\circ\top}_i(\boldsymbol{x}_0 - \boldsymbol{d}_j)\}$. By Lemma I.5 of Ref. [70]

$$\mathcal{C}_k := \text{conv}\left(\sum_{j=1}^{n} \tfrac{\partial \varphi_k}{\partial z_j}(\gamma(\boldsymbol{x}_0))\{\boldsymbol{v}^\circ_\ell : \ell \in I_j(\boldsymbol{x}_0)\}\right)$$

and so $\mathcal{C}_k$ is also a polytope for every $k \in K(\boldsymbol{x}_0)$. This implies that $\text{conv}\left(\bigcup_{k\in K(\boldsymbol{x}_0)} \mathcal{C}_k\right)$ is itself a polytope and hence in principle one can solve the above decision problem by linear programming.

In the next section we discuss easy algorithms to solve directly the decision problems corresponding to the previous pair of examples for $s = 2$.

4.6. Checking optimality in the planar case

In this section we discuss two related membership problems in $\mathbb{R}^2$. The first problem is to decide whether $\mathbf{0}$ belongs to the convex hull of a finite number of points. Moreover, the second problem is to find out whether $\mathbf{0}$ belongs to the convex hull of a finite number of different ℓ_q-circles.

The previous section shows that these decision problems are equivalent to testing whether a given point is an optimal solution of a min-max single facility continuous location problem with polyhedral gauges or ℓ_p-norms. Also in this section a description of a practical linear-time algorithm to solve the first decision problem together with its proof of correctness is given in Subsection 4.6.1. In Subsection 4.6.2 it is shown that the second decision problem can be reduced to the first.

Also in the same subsection the properties of the associated reduction algorithm are discussed.

Since in each step the optimality check has to be executed, it is important that an efficient algorithm is available to evaluate this decision.

A *global* optimality check is equivalent to the decision whether $\mathbf{0}$ belongs or not to $\text{conv}(\{\boldsymbol{p}_1, \ldots, \boldsymbol{p}_{\hat{\kappa}}\})$ with $\boldsymbol{p}_1, \ldots, \boldsymbol{p}_{\hat{\kappa}}$ some finite number of points in $\mathbb{R}^2$.

Moreover, if for each demand point $\boldsymbol{d}_i$, $1 \leq i \leq n$, the distance in (L) is given by some ℓ_{p_i}-norm $\boldsymbol{x} \longmapsto \|\boldsymbol{x}\|_{p_i}$, $1 < p_i < \infty$, then a *global* optimality check reduces for some $\hat{\kappa} \leq \kappa$ either to the decision whether $\mathbf{0}$ belongs to $\text{conv}\left(\bigcup_{1 \leq i \leq \hat{\kappa}} \boldsymbol{c}_i + r_i \mathcal{G}_{q_i}\right)$ with $\mathcal{G}_{q_i} := \{\boldsymbol{x} \in \mathbb{R}^2 : \|\boldsymbol{x}\|_{q_i} \leq 1\}$ the so-called unit ℓ_{q_i}-circle, $\boldsymbol{c}_i \in \mathbb{R}^s$, $r_i > 0$, $\frac{1}{p_i} + \frac{1}{q_i} = 1$ or to the decision whether $\mathbf{0}$ belongs to $\text{conv}(\{\boldsymbol{p}_1, \ldots, \boldsymbol{p}_{\hat{\kappa}}\})$.

Usually, in the last two problems $\hat{\kappa}$ is much smaller than κ. We call the first decision problem (D) and the second (D'). The next subsections describe some practical algorithms for solving (D) and (D') in $\mathbb{R}^2$. Observe that although (D') is only solved approximately, the solution can be found up to any degree of approximation.

4.6.1. Solving (D)

Clearly, if $p_i = 0$ for any $1 \leq i \leq \hat{\kappa}$ then 0 belongs to $\mathrm{conv}(\{p_1, p_2, \ldots, p_{\hat{\kappa}}\})$ and so we assume that $0 \notin \{p_1, p_2, \ldots, p_{\hat{\kappa}}\}$.

A straightforward way to check whether 0 belongs to the convex hull of a set of points $\{p_1, p_2, \ldots, p_{\hat{\kappa}}\}$ starts by constructing the convex hull of the enlarged set of points $\{0, p_1, p_2, \ldots, p_{\hat{\kappa}}\}$. This can be done using an efficient algorithm like Graham's scan, see Ref. [86]. After executing this $\mathcal{O}(\hat{\kappa} \log \hat{\kappa})$ algorithm we obtain the extreme points of the polytope $\mathrm{conv}(\{0, p_1, p_2, \ldots, p_{\hat{\kappa}}\})$. It can easily be verified that 0 is an extreme point of $\mathrm{conv}(\{0, p_1, p_2, \ldots, p_{\hat{\kappa}}\})$ if and only if 0 does not belong to $\mathrm{conv}(\{p_1, p_2, \ldots, p_{\hat{\kappa}}\})$. Although it is not difficult to adapt Graham's scan in such a way that the algorithm stops with a positive or a negative answer before the complete construction of the polytope $\mathrm{conv}(\{0, p_1, p_2, \ldots, p_{\hat{\kappa}}\})$, it still may take $\mathcal{O}(\hat{\kappa} \log \hat{\kappa})$ operations in the worst case. Since we have to solve this decision problem in each step of an iterative procedure, constructing an algorithm with lower complexity may reduce the overall computational effort. Observe that the best we can hope for is a linear time algorithm.

This may be achieved by means of the following construction. First, reformulate the problem into a linear programming problem (Ref. [57]) with 2 variables and $\hat{\kappa}$ constraints and then apply a known linear time algorithm, for instance one of those proposed in Ref. [28, 72, 94, 95]. However, although being linear in the number of constraints when the number of variables is fixed, the existing procedures exhibit exponential dependence on the space dimension. This dependence is at least of $\mathcal{O}\left(3^{s^2}\right)$ (Ref. [95]) which even for the simpler case $s = 2$ of our planar problem yields a solution procedure consisting of about $3^4 \hat{\kappa}$ operations. Although undoubtedly linear this is not attractive.

Before presenting a much more efficient algorithm for directly solving our decision problem in the planar case we need the following well-known definition and results from computational geometry (Ref. [86]).

Definition 4.6.1. *A point $p_0 \in \mathbb{R}^2$ is to the right of a directed segment from point p_1 to point p_2 if it is not an element of the line going through p_1 and p_2 and if moving along this line in the direction of p_1 to p_2 the point p_0 is an element of the* *right hand side half-plane.*

The definition of points to the left of a directed segment is similar. Moreover, given a line segment from $\boldsymbol{p}_1$ to $\boldsymbol{p}_2$ and another line segment from $\boldsymbol{p}_3$ to $\boldsymbol{p}_4$, a point $\boldsymbol{p}_0$ is said to be *between* these two line segments when $\boldsymbol{p}_0$ is to the right of one and to the left of the other.

In the remainder, a directed line segment from $\boldsymbol{p}_i$ to $\boldsymbol{p}_j$ is denoted by $[\boldsymbol{p}_i, \boldsymbol{p}_j]$.

The following result can be used in order to determine to which side of a directed line segment a given point belongs to.

Lemma 4.6.1. *Let* $\boldsymbol{p}_i = (p_{i_1}, p_{i_2})$, $0 \leq i \leq 2$. *If*

$$\Delta := \det\left(\begin{bmatrix} p_{0_1} & p_{0_2} & 1 \\ p_{1_1} & p_{1_2} & 1 \\ p_{2_1} & p_{2_2} & 1 \end{bmatrix}\right)$$

then the next results hold.

1. $\boldsymbol{p}_0$ *is colinear with* $[\boldsymbol{p}_1, \boldsymbol{p}_2]$ *if and only if* $\Delta = 0$;
2. $\boldsymbol{p}_0$ *is to the right of* $[\boldsymbol{p}_1, \boldsymbol{p}_2]$ *if and only if* $\Delta < 0$;
3. $\boldsymbol{p}_0$ *is to the left of* $[\boldsymbol{p}_1, \boldsymbol{p}_2]$ *if and only if* $\Delta > 0$.

Proof: A proof of this result can be found in Ref. [86] and hence it is omitted. □

In order to explain our linear time algorithm we also need the following result which is a trivial consequence of Lemma 4.6.1.

Lemma 4.6.2. *The point* $\boldsymbol{p}_0$ *is between the two directed line segments* $[\boldsymbol{p}_1, \boldsymbol{p}_2]$ *and* $[\boldsymbol{p}_3, \boldsymbol{p}_4]$ *if and only if* Δ_1 *and* Δ_2 *have opposite signs where*

$$\Delta_1 := \det\left(\begin{bmatrix} p_{0_1} & p_{0_2} & 1 \\ p_{1_1} & p_{1_2} & 1 \\ p_{2_1} & p_{2_2} & 1 \end{bmatrix}\right) \quad \text{and} \quad \Delta_2 := \det\left(\begin{bmatrix} p_{0_1} & p_{0_2} & 1 \\ p_{3_1} & p_{3_2} & 1 \\ p_{4_1} & p_{4_2} & 1 \end{bmatrix}\right).$$

Observe that in the sequel of this discussion $\boldsymbol{p}_0$ is always $\mathbf{0}$. Therefore, only a 2×2 determinant is actually computed.

Let us denote by $\mathcal{C}(\boldsymbol{l}, \boldsymbol{r})$ the closed convex cone with apex at $\mathbf{0}$ and generated by the line segments $[\mathbf{0}, \boldsymbol{l}]$ and $[\mathbf{0}, \boldsymbol{r}]$, and let $\boldsymbol{l}$ and $\boldsymbol{r}$ be chosen in such a way that $\boldsymbol{l}$ is to the left of $[\mathbf{0}, \boldsymbol{r}]$. The test $\boldsymbol{x} \in \mathcal{C}(\boldsymbol{l}, \boldsymbol{r})$ can be implemented using Lemma 4.6.2.

Input: $\hat{\kappa}$ points $\boldsymbol{p}_1, \ldots, \boldsymbol{p}_{\hat{\kappa}} \in \mathbf{R}^2$
Output: *YES* (if $0 \in \operatorname{conv}(\{\boldsymbol{p}_1, \ldots, \boldsymbol{p}_{\hat{\kappa}}\})$) or *NO* (otherwise)
Step 1: *if* $\hat{\kappa} = 1$ *then if* $\boldsymbol{p}_1 = \mathbf{0}$ *then* stop with *YES*
else stop with *NO*
else i := 2;
proceed to **Step 2**
Step 2: *if* $\boldsymbol{p}_i$ is colinear with $[\mathbf{0}, \boldsymbol{p}_1]$
then if $\mathbf{0} \in [\boldsymbol{p}_1, \boldsymbol{p}_i]$ *then* stop with *YES*
else $i := i + 1$;
if $i \leq \hat{\kappa}$ *then* repeat **Step 2**
else stop with *NO*
else if $\boldsymbol{p}_i$ is to the right of $[\mathbf{0}, \boldsymbol{p}_1]$
then $\boldsymbol{l} := \boldsymbol{p}_1$; $\boldsymbol{r} := \boldsymbol{p}_i$
else $\boldsymbol{r} := \boldsymbol{p}_1$; $\boldsymbol{l} := \boldsymbol{p}_i$;
$i := i + 1$;
if $i \leq \hat{\kappa}$ *then* proceed to **Step 3**
else stop with *NO*
Step 3: *if* $\boldsymbol{p}_i \in \mathcal{C}(-\boldsymbol{r}, \boldsymbol{l})$ *then* $\boldsymbol{l} := \boldsymbol{p}_i$
else if $\boldsymbol{p}_i \in \mathcal{C}(\boldsymbol{r}, -\boldsymbol{l})$ *then* $\boldsymbol{r} := \boldsymbol{p}_i$
else if $\boldsymbol{p}_i \in \mathcal{C}(-\boldsymbol{l}, -\boldsymbol{r})$
then stop with *YES*;
$i := i + 1$;
if $i \leq \hat{\kappa}$ *then* repeat **Step 3**
else stop with *NO*

Algorithm 4.2: The decision algorithm

Our algorithm is presented in Algorithm 4.2. This algorithm is a simplification of the algorithm proposed in Ref. [36] and first presented in Ref. [41].

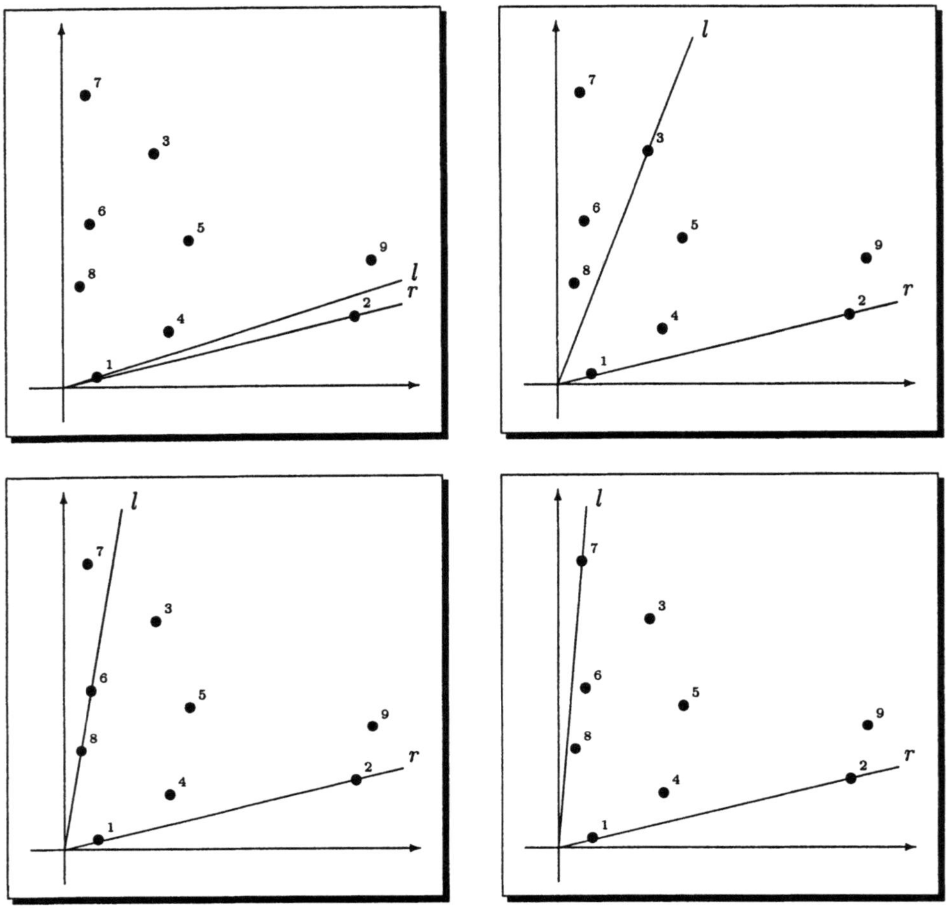

Figure 4.10: Example of execution of the decision algorithm

First, the trivial case of just one point is eliminated in **Step 1**. If there are two or more points then an initialization occurs in **Step 2**. For the purpose of initialization $\boldsymbol{p}_1$ is selected and the algorithm searches for the first point in $\{\boldsymbol{p}_2, \ldots, \boldsymbol{p}_{\hat{k}}\}$ noncolinear with $[\mathbf{0}, \boldsymbol{p}_1]$. If during the search $\mathbf{0}$ is found to belong to $[\boldsymbol{p}_1, \boldsymbol{p}_i]$ for any i the algorithm stops without reaching the main loop and answers *YES*.

However, if all the points are exhausted without finding either a noncolinear point or a point such that $\mathbf{0} \in [\boldsymbol{p}_1, \boldsymbol{p}_i]$, the algorithm also stops before the main loop but answers *NO*. If a noncolinear $\boldsymbol{p}_i$ is found then $\boldsymbol{l}$ (left) and $\boldsymbol{r}$ (right) are assigned to $\boldsymbol{p}_1$ and $\boldsymbol{p}_i$ or vice versa and the main loop begins in **Step 3**.

The main loop is based on the trivial observation that $\mathbf{0}$ is outside $\text{conv}(\{\boldsymbol{p}_1, \ldots, \boldsymbol{p}_{\hat{\kappa}}\})$ if and only if some $\boldsymbol{l}$ and $\boldsymbol{r}$ exist in $\{\boldsymbol{p}_1, \ldots, \boldsymbol{p}_{\hat{\kappa}}\}$ defining a closed convex cone $C(\boldsymbol{l}, \boldsymbol{r})$ such that $\text{conv}(\{\boldsymbol{p}_1, \ldots, \boldsymbol{p}_{\hat{\kappa}}\}) \subseteq C(\boldsymbol{l}, \boldsymbol{r})$.

The main loop (**Step 3**) updates $C(\boldsymbol{l}, \boldsymbol{r})$ while it remains convex or concludes that $\mathbf{0} \in \text{conv}(\{\boldsymbol{p}_1, \ldots, \boldsymbol{p}_{\hat{\kappa}}\})$. In each iteration the plane is partitioned by the interior-disjoint cones $C(\boldsymbol{l}, \boldsymbol{r})$, $C(-\boldsymbol{r}, \boldsymbol{l})$, $C(\boldsymbol{r}, -\boldsymbol{l})$ and $C(-\boldsymbol{l}, -\boldsymbol{r})$ and one of the following actions is carried out. In case $\boldsymbol{p}_i \in C(\boldsymbol{l}, \boldsymbol{r})$ nothing is done. However, if $\boldsymbol{p}_i \in C(-\boldsymbol{r}, \boldsymbol{l})$ then $\boldsymbol{l}$ is replaced by $\boldsymbol{p}_i$, while if $\boldsymbol{p}_i \in C(\boldsymbol{r}, -\boldsymbol{l})$ then $\boldsymbol{r}$ is replaced by $\boldsymbol{p}_i$. Finally, if $\boldsymbol{p}_i \in C(-\boldsymbol{r}, -\boldsymbol{l})$ then the algorithm stops with *YES* since $\mathbf{0}$ belongs to the triangle $\text{conv}(\{\boldsymbol{l}, \boldsymbol{r}, \boldsymbol{p}_i\})$ and this is a subset of $\text{conv}(\{\boldsymbol{p}_1, \ldots, \boldsymbol{p}_{\hat{\kappa}}\})$.

Figure 4.10 shows an example of the execution of Algorithm 4.2 to decide if $\mathbf{0}$ belongs to the convex hull of 9 given points. The first picture in Figure 4.10 shows the result of **Step 2**, initializing $\boldsymbol{l} = \boldsymbol{p}_1$ and $\boldsymbol{r} = \boldsymbol{p}_2$. The remaining three pictures show the steps in which $C(\boldsymbol{l}, \boldsymbol{r})$ is updated for the example, namely $i = 3$, $i = 6$ and $i = 7$. The last iteration is reached in this example, $i = 9$, and $C(\boldsymbol{l}, \boldsymbol{r})$ stays convex, therefore the decision in this case is *NO*, $\mathbf{0}$ does not belong to the convex hull of the union of the given points.

Algorithm 4.2 is thus a very simple and correct algorithm. Moreover, as the following result states, both the total number of operations and storage requirements are obviously linear functions in the number of points with very small coefficients. Actually, the total number of logical conditions evaluated is below $4\hat{\kappa}$ while the total number of arithmetic operations is below $7\hat{\kappa}$, as the proof of the next lemma shows.

Lemma 4.6.3. *Algorithm 4.2 gives the correct answer in all cases. Moreover, both its computational complexity and storage requirements are of* $\mathcal{O}(\hat{\kappa})$.

Proof: The correctness of the algorithm follows from the previous discussion. The storage requirements amount to storing $\boldsymbol{p}_1, \ldots, \boldsymbol{p}_{\hat{\kappa}}$ plus a pair of index pointers to identify $\boldsymbol{l}$ and $\boldsymbol{r}$. For each point $\boldsymbol{p}_i$, $1 \leq i \leq \hat{\kappa}$, one (**Step 2** by means of Lemma 4.6.1) or two (**Step 3** by means of Lemma 4.6.2) 2×2 determinants are computed and at most four logical conditions are evaluated. Each 2×2 determinant requires two multiplications and one subtraction. This yields the complexity as stated in the lemma. □

Observe that Algorithm 4.2 solves the slightly more general decision problem: given $\boldsymbol{p}_0$ and $\{\boldsymbol{p}_1, \ldots, \boldsymbol{p}_{\hat{\kappa}}\}$ is $\boldsymbol{p}_0 \in \operatorname{conv}(\{\boldsymbol{p}_1, \ldots, \boldsymbol{p}_{\hat{\kappa}}\})$? In fact, to convert this to (D) by translation, it is enough to replace each $\boldsymbol{p}_i$, $0 \leq i \leq \hat{\kappa}$, by $\boldsymbol{p}_i - \boldsymbol{p}_0$.

In the next subsection we show how the decision problem (D') can be converted into the decision problem (D) and present a numerical procedure to carry out this reduction.

4.6.2. Solving (D')

In the second decision problem we want to verify whether $\mathbf{0}$ belongs to

$$\operatorname{conv}\left(\bigcup_{1 \leq i \leq \hat{\kappa}} \boldsymbol{c}_i + r_i \mathcal{G}_{q_i}\right).$$

Clearly, by computing $\|\boldsymbol{c}_i\|_{q_i}$, we first check whether $\mathbf{0} \in \boldsymbol{c}_i + r_i \mathcal{G}_{q_i}$, i.e. $\|\boldsymbol{c}_i\|_{q_i} \leq r_i$, for some $1 \leq i \leq \hat{\kappa}$. If this holds, then $\mathbf{0}$ clearly belongs to $\operatorname{conv}\left(\bigcup_{1 \leq i \leq \hat{\kappa}} \boldsymbol{c}_i + r_i \mathcal{G}_{q_i}\right)$. However, if this is not true, we have to solve the following nontrivial decision problem

$$0 \notin \boldsymbol{c}_i + r_i \mathcal{G}_{q_i},\ 1 \leq i \leq \hat{\kappa} \overset{?}{\Rightarrow} \mathbf{0} \in \operatorname{conv}\left(\bigcup_{1 \leq i \leq \hat{\kappa}} \boldsymbol{c}_i + r_i \mathcal{G}_{q_i}\right).$$

For $\mathbf{0} \notin \boldsymbol{c}_i + r_i \mathcal{G}_{q_i}$ there must exist one pair of supporting hyperplanes $\mathcal{H}_i^L$ and $\mathcal{H}_i^R$ of $\boldsymbol{c}_i + r_i \mathcal{G}_{q_i}$ going through $\mathbf{0}$. Moreover, let $\boldsymbol{t}_i^L$ and $\boldsymbol{t}_i^R$ denote the pair of unique intersection, or tangent, points of the corresponding hyperplanes and $\boldsymbol{c}_i + r_i \mathcal{G}_{q_i}$. We refer to $\mathcal{H}_i^L$ as the *left* tangent hyperplane and to $\mathcal{H}_i^R$ as the *right* tangent hyperplane. The left hyperplane is defined as the one whose tangent point is to the left of $[\mathbf{0}, \boldsymbol{c}_i]$ while the right hyperplane is given by the other one.

Keeping in mind the definition of $\boldsymbol{t}_i^L$ and $\boldsymbol{t}_i^R$ one can now prove the following result for (D').

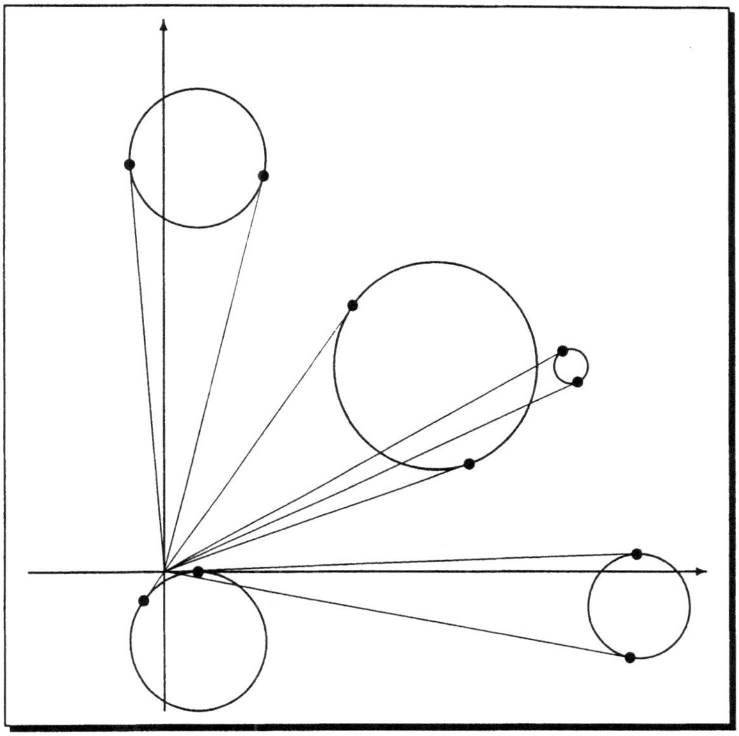

Figure 4.11: Interpretation of the conversion result

Lemma 4.6.4. *If* $\mathbf{0}$ *does not belong to* $\mathbf{c}_i + r_i\mathcal{G}_{q_i}$ *for every* $1 \le i \le \hat{\kappa}$ *then*

$$\mathbf{0} \in \operatorname{conv}\left(\bigcup_{1\le i\le\hat{\kappa}} \mathbf{c}_i + r_i\mathcal{G}_{q_i}\right)$$

if and only if

$$\mathbf{0} \in \operatorname{conv}(\{\boldsymbol{t}_1^L, \boldsymbol{t}_1^R, \ldots, \boldsymbol{t}_{\hat{\kappa}}^L, \boldsymbol{t}_{\hat{\kappa}}^R\}).$$

Proof: Since $\boldsymbol{t}_i^L, \boldsymbol{t}_i^R \in \mathbf{c}_i + r_i\mathcal{G}_{q_i}$ for every $1 \le i \le \hat{\kappa}$ it follows that

$$\operatorname{conv}(\{\boldsymbol{t}_1^L, \boldsymbol{t}_1^R, \ldots, \boldsymbol{t}_{\hat{\kappa}}^L, \boldsymbol{t}_{\hat{\kappa}}^R\}) \subseteq \operatorname{conv}\left(\bigcup_{1\le i\le\hat{\kappa}} \mathbf{c}_i + r_i\mathcal{G}_{q_i}\right)$$

and hence the second condition implies the first. For the proof of the reverse implication let $\mathbf{0} \in \operatorname{conv}\left(\bigcup_{1\le i\le\hat{\kappa}} \mathbf{c}_i + r_i\mathcal{G}_{q_i}\right)$, i.e. there exist some $\alpha_i \ge 0$, $1 \le i \le \hat{\kappa}$, such that $\sum_{i=1}^{\hat{\kappa}} \alpha_i = 1$ and $\mathbf{0} = \sum_{i=1}^{\hat{\kappa}} \alpha_i \boldsymbol{p}_i$, with each $\boldsymbol{p}_i$ belonging to $\boldsymbol{c}_i + r_i \mathcal{G}_{q_i}$, $1 \leq i \leq \hat{\kappa}$. Consider now for each $\boldsymbol{p}_i$ belonging to $\boldsymbol{c}_i + r_i \mathcal{G}_{q_i}$ the straight line $\mathcal{K}_i$ going through $\mathbf{0}$ and $\boldsymbol{p}_i$. Since $\boldsymbol{p}_i$ belongs to the cone generated by the two hyperplanes $\mathcal{H}_i^L$ and $\mathcal{H}_i^R$ it follows that the straight line $\mathcal{K}_i$ intersects the line segment connecting $\boldsymbol{t}_i^L$ and $\boldsymbol{t}_i^R$ in the point $\boldsymbol{q}_i$. Hence, we can find constants $c_i > 0$ and $0 \leq \tau_i \leq 1$ such that $\boldsymbol{p}_i = c_i \boldsymbol{q}_i$ and $\boldsymbol{q}_i = \tau_i \boldsymbol{t}_i^L + (1-\tau_i)\boldsymbol{t}_i^R$. This implies that $\boldsymbol{p}_i = c_i\left(\tau_i \boldsymbol{t}_i^L + (1-\tau_i)\boldsymbol{t}_i^R\right)$ and so, after normalizing, it follows that $\mathbf{0}$ can be written as a convex combination of the tangent points $\boldsymbol{t}_1^L, \boldsymbol{t}_1^R, \ldots, \boldsymbol{t}_{\hat{\kappa}}^L, \boldsymbol{t}_{\hat{\kappa}}^R$. □

Figure 4.11 shows five instances of the ℓ_{q_i}-circles and the five corresponding pairs of tangent points. Observe, for the sake of illustration, that in Figure 4.11 Euclidean (ℓ_2) circles are used to represent general ℓ_{q_i}-circles. We keep this practice throughout this section but, as an aid to visualising general ℓ_{q_i}-circles, we provide the next figure.

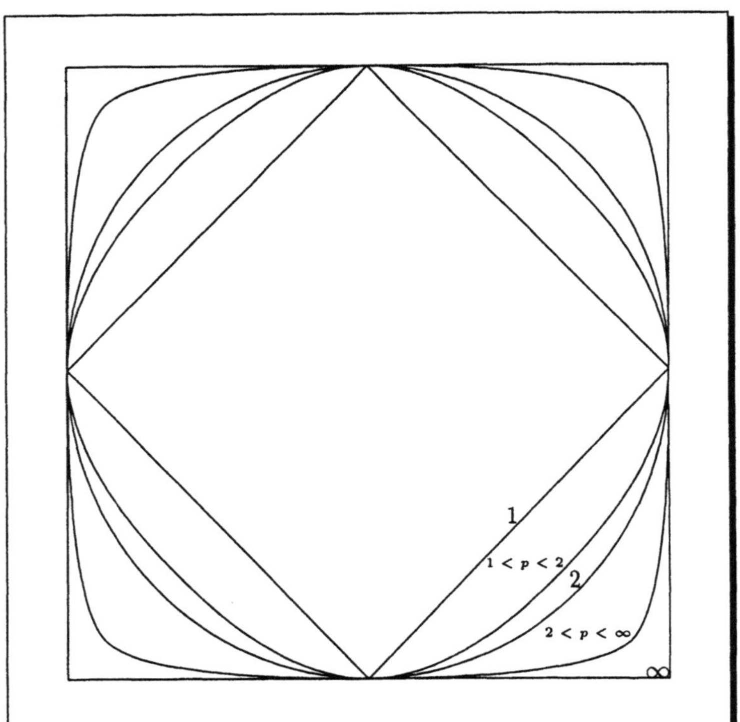

Figure 4.12: The family of ℓ_p-circles, $1 \leq p \leq \infty$

The result stated by Lemma 4.6.4 gives us the possibility to reduce every instance of the decision problem (D') into an instance of the decision problem (D). First, the normal vectors of the tangent hyperplanes need to be computed. After that we determine the tangent point of each hyperplane and apply Algorithm 4.2 to the set of these points. Let us denote as in Lemma 4.6.4 these points by $\{t_1^L, t_1^R, \dots, t_{\bar{K}}^L, t_{\bar{K}}^R\}$.

In the remainder we omit the indices for the sake of notational convenience and so a tangent hyperplane is given by $\mathcal{H}$ regardless of being left or right. We discuss now the computation of each $a \in \mathbb{R}^2$ with a a normal vector of $\mathcal{H}$, i.e.

$$\mathcal{H} = \{z \in \mathbb{R}^2 : a^\top z = 0\}.$$

Let us assume without loss of generality that $\|a\|_p = 1$ with p such that $\frac{1}{p} + \frac{1}{q} = 1$ and $a^\top x \leq 0$ holds for every $x \in c + r\mathcal{G}_q$. This means that a points from 0 to the halfspace not containing $c + r\mathcal{G}_q$.

Since there is a tangent point $t \in \mathcal{H} \cap (c + r\mathcal{G}_q)$ it must follow that $a^\top t = 0$ and so by the previous condition we obtain

$$\max\{a^\top x : x \in c + r\mathcal{G}_q\} = 0. \tag{4.2}$$

Clearly, every element $x \in c + r\mathcal{G}_q$ can be written as $x = c + ry$ where $y \in \mathcal{G}_q$ and hence (4.2) is equivalent to

$$a^\top c + r \max\{a^\top y : y \in \mathcal{G}_q\} = 0. \tag{4.3}$$

Applying Hölder's inequality (Ref. [106]) to the second term in (4.3) it follows that

$$\max\{a^\top y : y \in \mathcal{G}_q\} = \|a\|_p. \tag{4.4}$$

Hence, by (4.3) and (4.4) we obtain that the normal vector a of $\mathcal{H}$ much satisfy $\|a\|_p = -\frac{1}{r}a^\top c > 0$ and so, using the normalization $\|a\|_p = 1$ and defining $c' := -\frac{1}{r}c$, yields $c'^\top a = 1$.

By the above observations we have verified that the vector a satisfies

$$\begin{cases} c'^\top a & = 1 \\ \|a\|_p & = 1 \end{cases}. \tag{4.5}$$

A geometrical argument based on the existence of exactly two supporting hyperplanes guarantees that there are exactly two real valued vectors solving this system, corresponding to the normal vectors of these hyperplanes. Since it is not yet possible to decide which one is left or right we index them as *one* and *two*.

Unfortunately, in general it is not possible to write down analytical solutions of these nonlinear equations and so we have to use a numerical procedure to find them. However, for some special cases this can be done. These cases are listed now.

We denote by $\boldsymbol{e}_i$ the ith unit vector, that is $\boldsymbol{e}_1 = (1,0)$ and $\boldsymbol{e}_2 = (0,1)$.

1. Particular values of p (or q)

   (a) $p = 1$ $(q = \infty)$ $\boldsymbol{a}_j = \begin{cases} \left(\frac{1-c_2'}{c_1'-c_2'}, \frac{c_1'-1}{c_1'-c_2'}\right) & \text{if it belongs to } \mathcal{C}(\boldsymbol{e}_1, \boldsymbol{e}_2) \\ \left(\frac{1-c_2'}{c_1'+c_2'}, \frac{c_1'+1}{c_1'+c_2'}\right) & \text{if it belongs to } \mathcal{C}(\boldsymbol{e}_2, -\boldsymbol{e}_1) \\ \left(\frac{1+c_2'}{c_1'-c_2'}, \frac{-1-c_1'}{c_1'-c_2'}\right) & \text{if it belongs to } \mathcal{C}(-\boldsymbol{e}_1, -\boldsymbol{e}_2) \\ \left(\frac{1+c_2'}{c_1'+c_2'}, \frac{1-c_1'}{c_1'+c_2'}\right) & \text{if it belongs to } \mathcal{C}(-\boldsymbol{e}_2, \boldsymbol{e}_1) \end{cases}$;

   (b) $p = 2$ $(q = 2)$ $\boldsymbol{a}_j = \left(\frac{c_1' \pm c_2'\sqrt{\Delta'}}{\Delta'+1}, \frac{c_2' \mp c_1'\sqrt{\Delta'}}{\Delta'+1}\right)$, where $\Delta' = {c_1'}^2 + {c_2'}^2 - 1$;

   (c) $p = \infty$ $(q = 1)$ $\boldsymbol{a}_j = \begin{cases} \left(\frac{1-c_2'}{c_1'}, 1\right) & \text{if } -1 < \frac{1-c_2'}{c_1'} < 1 \\ \left(-1, \frac{1+c_1'}{c_2'}\right) & \text{if } -1 < \frac{1+c_1'}{c_2'} < 1 \\ \left(\frac{1+c_2'}{c_1'}, -1\right) & \text{if } -1 < \frac{1+c_2'}{c_1'} < 1 \\ \left(1, \frac{1-c_1'}{c_2'}\right) & \text{if } -1 < \frac{1-c_1'}{c_2'} < 1 \end{cases}$;

2. Particular values of $\boldsymbol{c}'$

   (a) $c_1' = 0 \Rightarrow \boldsymbol{a}_j = \left(\pm\frac{(|c_2'|^p-1)^{1/p}}{|c_2'|}, \frac{1}{c_2'}\right)$;

   (b) $c_2' = 0 \Rightarrow \boldsymbol{a}_j = \left(\frac{1}{c_1'}, \pm\frac{(|c_1'|^p-1)^{1/p}}{|c_1'|}\right)$;

   (c) incomplete solutions for particular values of $\boldsymbol{c}'$

      i. $c_1' = 1 \Rightarrow \boldsymbol{a}_1 = (1,0)$;

      ii. $c_1' = -1 \Rightarrow \boldsymbol{a}_1 = (-1,0)$;

      iii. $c_2' = 1 \Rightarrow \boldsymbol{a}_1 = (0,1)$;

      iv. $c_2' = -1 \Rightarrow \boldsymbol{a}_1 = (0,-1)$.

Note that for (2c) only one solution was found analytically. Cases (1a) and (1c) are only listed here for completeness since $1 < p < \infty$ is assumed. In fact, the cases $p = 1$ and $p = \infty$ correspond to polyhedral gauges. Moreover, the solution given in (1a) is needed at a further point.

In order to find a solution for the other cases we need to use a nonlinear procedure. Without loss of generality assume now that $c'_1, c'_2 \notin \{-1, 0, 1\}$.

Observe that the solution of (4.5) is given by the two points where the line $\mathcal{K} := \{\boldsymbol{x} \in \mathbb{R}^2 : \boldsymbol{c}'^{\mathsf{T}}\boldsymbol{x} = 1\}$ intersects the unit ℓ_p-circle $\mathcal{G}_p$.

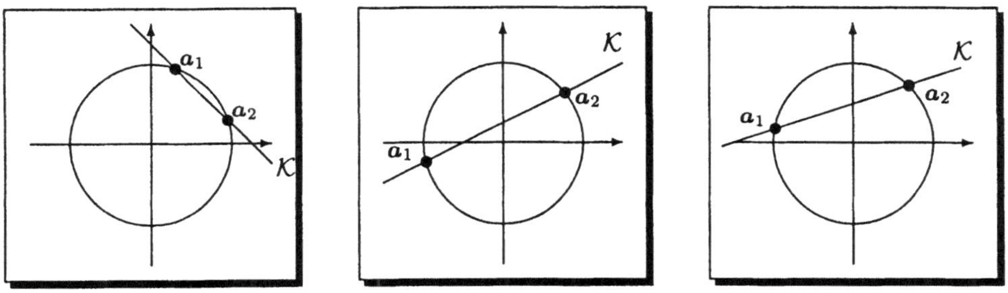

Figure 4.13: Intersections of $\mathcal{K}$ and the unit ℓ_p-circle

In Figure 4.13 the three cases modulo possible rotations are shown. Clearly, every point $\boldsymbol{x} \in \mathcal{K}$ can be represented by $\boldsymbol{x} = \boldsymbol{b} + \mu\boldsymbol{u}$ where $\boldsymbol{b}$ is some point on $\mathcal{K}$ and $\boldsymbol{u}$ is a vector parallel to $\mathcal{K}$.

The problem now reduces to finding the roots of the equation $\psi(\mu) = 1$ with the function $\psi : \mathbb{R} \longrightarrow \mathbb{R}$ given by

$$\psi(\mu) := \|\boldsymbol{b} + \mu\boldsymbol{u}\|_p.$$

We first give some properties of the function ψ. These properties are independent of the choice of $\boldsymbol{b}$ and $\boldsymbol{u}$.

Lemma 4.6.5. *The real function $\psi : \mu \longmapsto \|\boldsymbol{b} + \mu\boldsymbol{u}\|_p$ is a differentiable convex function. Moreover, it is Lipschitz continuous with Lipschitz constant $\|\boldsymbol{u}\|_p$.*

Proof: The convexity of the function ψ is an easy consequence of the convexity of the mapping $\boldsymbol{v} \longmapsto \|\boldsymbol{v}\|_p$. Also, since this mapping is Frechet differentiable in

$\mathbb{R}^2 \setminus \{0\}$ and $\mathbf{0}$ does not belong to the line $\mathcal{K}$, the differentiability of ψ follows. Hence, it only remains to verify the Lipschitz continuity of ψ. Clearly, for every μ_1 and μ_2 we obtain

$$\begin{aligned} |\psi(\mu_1) - \psi(\mu_2)| &= \Big| \|\boldsymbol{b} + \mu_1 \boldsymbol{u}\|_p - \|\boldsymbol{b} + \mu_2 \boldsymbol{u}\|_p \Big| \\ &\leq \|(\boldsymbol{b} + \mu_1 \boldsymbol{u}) - (\boldsymbol{b} + \mu_2 \boldsymbol{u})\|_p \\ &= \|(\mu_1 - \mu_2)\boldsymbol{u}\|_p = \|\boldsymbol{u}\|_p \Big|\mu_1 - \mu_2\Big| \end{aligned}$$

and this concludes the proof. □

For the computation of the two roots μ_1' and μ_2', corresponding to the normal vectors $\boldsymbol{a}_1 := \boldsymbol{b} + \mu_1' \boldsymbol{u}$ and $\boldsymbol{a}_2 := \boldsymbol{b} + \mu_2' \boldsymbol{u}$, we first derive an interval containing both roots. Recall the well-known inequality (Ref. [106])

$$1 \leq p_1 < p_2 \leq \infty \;\Rightarrow\; \|\boldsymbol{x}\|_{p_1} \geq \|\boldsymbol{x}\|_{p_2}.$$

Moreover, if $\boldsymbol{x}$ does not belong to any of the coordinate axes it is easy to show that the inequality sign between the ℓ_p-norms can be replaced by a strict inequality sign. From this inequality it follows that the unit ℓ_p-circle is contained in the unit ℓ_∞-circle. Hence (see also Figure 4.14) the two intersection points $\boldsymbol{\alpha}_1$ and $\boldsymbol{\alpha}_2$ of the line $\mathcal{K}$ and the unit ℓ_∞-circle are outside the unit ℓ_p-circle and so it must follow that both solutions $\boldsymbol{a}_1$ and $\boldsymbol{a}_2$ of (4.5) belong to the open line segment $]\boldsymbol{\alpha}_1, \boldsymbol{\alpha}_2[$.

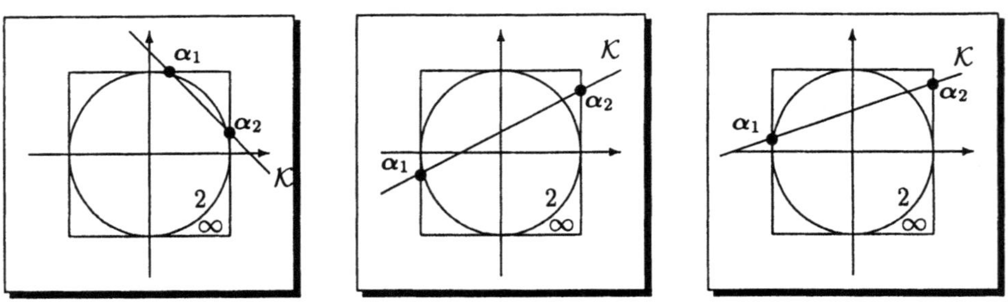

Figure 4.14: Starting points

Since by case (1a) $\boldsymbol{\alpha}_1$ and $\boldsymbol{\alpha}_2$ can be computed analytically, we take $\boldsymbol{b} := \boldsymbol{\alpha}_1$ and $\boldsymbol{u} := \boldsymbol{\alpha}_2 - \boldsymbol{\alpha}_1$ (see Figure 4.15).

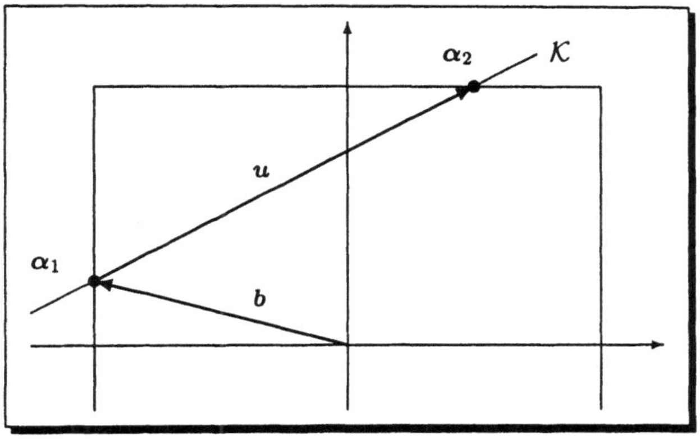

Figure 4.15: Construction of ψ

This implies that the pair of different roots μ_1' and μ_2' belongs to $]0,1[$. Also, since $\psi(\mu_1') = \psi(\mu_2') = 1$ and $\psi(0) > 1$, $\psi(1) > 1$ it follows by the convexity of ψ that a minimum point belongs to $]\mu_1', \mu_2'[$ and so again by the convexity of ψ we obtain (see Figure 4.16) that $\psi'(\mu) < 0$ if $\mu \leq \mu_1'$ and $\psi'(\mu) > 0$ if $\mu \geq \mu_2'$.

These observations allow us to state the following well-known result for the Newton-Raphson method (Ref. [24]). We list an alternative short proof for the sake of completeness.

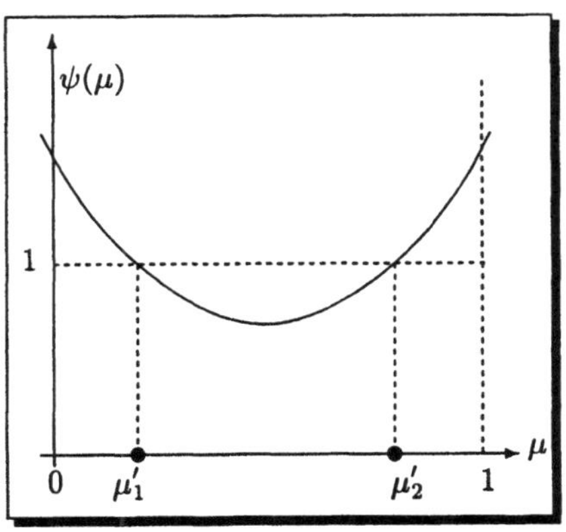

Figure 4.16: Sample graph of ψ

Lemma 4.6.6. *If the Newton-Raphson method is applied with starting point* 0 *to solve the equation* $\psi(\mu) - 1 = 0$ *it produces an increasing sequence of iterates converging from below to* μ_1'.

Proof: Let us denote by μ^ℓ, $\ell = 1, 2, \ldots$, the sequence of iterates produced by the Newton-Raphson method, i.e.

$$\mu^{\ell+1} := \mu^\ell + \frac{1 - \psi(\mu^\ell)}{\psi'(\mu^\ell)}. \tag{4.6}$$

By the subgradient inequality (Ref. [12]) it follows for every μ^ℓ that

$$1 = \psi(\mu_1') \geq \psi(\mu^\ell) + \psi'(\mu^\ell)(\mu_1' - \mu^\ell). \tag{4.7}$$

Take now $\mu^0 := 0$. Since $\psi'(\mu^0) < 0$ and $\psi(\mu^0) > 1$ we obtain

$$\frac{1 - \psi(\mu^0)}{\psi'(\mu^0)} > 0. \tag{4.8}$$

By (4.6) and (4.7) and taking $\ell = 0$ leads to

$$\mu^1 := \mu^0 + \frac{1 - \psi(\mu^0)}{\psi'(\mu^0)} \leq \mu_1'. \tag{4.9}$$

Now from (4.6), (4.8) and (4.9) we conclude that $\mu^0 < \mu^1 \leq \mu_1'$ and by induction it is easily proved that μ^ℓ is increasing and bounded from above by μ_1'. Due to $\mu^\ell \leq \mu_1'$ for every $\ell \geq 0$ it follows that the sequence converges and satisfies $\mu^\infty := \lim_{\ell \uparrow \infty} \mu^\ell \leq \mu_1'$. After observing that the derivative ψ' is continuous and negative on $[0, \mu_1']$ we obtain from (4.6) that this limit satisfies

$$\mu^\infty = \mu^\infty + \frac{1 - \psi(\mu^\infty)}{\psi'(\mu^\infty)}$$

and hence $\psi(\mu^\infty) = 1$. By the strict monotonicity of ψ on $[0, \mu_1']$ this finally yields $\mu^\infty = \mu_1'$. □

In order to guarantee finite termination it is now important to derive a stopping rule for this procedure meeting any prespecified error bound.

First we show that a minimum point of ψ, or equivalently a point μ' such that $\psi'(\mu') = 0$, is analytically computable.

We recall that ψ is differentiable everywhere. Moreover, its derivative ψ' has the following expression

$$\psi'(\mu) = \psi(\mu)^{1-p} \sum_{i=1}^{2} |b_i + \mu u_i|^{p-2}(b_i + \mu u_i)u_i. \tag{4.10}$$

Hence μ' satisfies the equation

$$\psi(\mu')^{1-p} \sum_{i=1}^{2} |b_i + \mu' u_i|^{p-2}(b_i + \mu' u_i)u_i = 0.$$

Since $\mathbf{0} \notin \mathcal{K}$ it follows that $\psi(\mu) > 0$ for every μ and so μ' must satisfy

$$|b_1 + \mu' u_1|^{p-2}(b_1 + \mu' u_1)u_1 = -|b_2 + \mu' u_2|^{p-2}(b_2 + \mu' u_2)u_2. \tag{4.11}$$

Obviously $\boldsymbol{u} \neq \mathbf{0}$ and so only the following two cases hold.

1. Either $u_1 = 0$ or $u_2 = 0$ but not both.

   Without loss of generality we can assume that $u_1 = 0$. In this case (4.11) reduces to $b_2 + \mu' u_2 = 0$ or

   $$\mu' = -\frac{b_2}{u_2}$$

   and $\psi(\mu') = |b_1|$.

2. Both $u_1 \neq 0$ and $u_2 \neq 0$.

   After taking absolute values in (4.11) we obtain

   $$|b_1 + \mu' u_1|^{p-1}|u_1| = |b_2 + \mu' u_2|^{p-1}|u_2|.$$

   Raising now both members to the power $1/(p-1)$ it follows that

   $$(b_1 + \mu' u_1) = \pm(b_2 + \mu' u_2)\left|\frac{u_2}{u_1}\right|^{1/(p-1)},$$

   and so we get

   $$\mu' = -\frac{b_1 \pm \left|\frac{u_2}{u_1}\right|^{1/(p-1)} b_2}{u_1 \pm \left|\frac{u_2}{u_1}\right|^{1/(p-1)} u_2}.$$

Substituting the above expression for μ' in (4.11) one can check that the undetermined sign $\pm$ should be a "+" if u_2 and u_1 have the same sign and a "−" otherwise.

This leads to the final expression where sign(x) stands for the sign function

$$\mu' = -\frac{b_1 + \operatorname{sign}\left(\frac{u_2}{u_1}\right)\left|\frac{u_2}{u_1}\right|^{1/(p-1)} b_2}{u_1 + \operatorname{sign}\left(\frac{u_2}{u_1}\right)\left|\frac{u_2}{u_1}\right|^{1/(p-1)} u_2}.$$

For this μ' it follows that $\psi(\mu') = |u_1 b_2 - u_2 b_1|/\|u\|_q$.

The next lemma provides a stopping rule for the Newton-Raphson procedure.

Lemma 4.6.7. *If the Newton-Raphson method is applied with starting point* 0 *to solve the equation* $\psi(\mu) - 1 = 0$ *it follows for every* $\ell \geq 1$ *that*

$$\mu^\ell \leq \mu_1' \leq \sigma^\ell$$

where

$$\sigma^\ell := \mu' + (\mu^\ell - \mu')\frac{1 - \psi(\mu')}{\psi(\mu^\ell) - \psi(\mu')}. \tag{4.12}$$

Moreover, the sequence σ^ℓ *is decreasing and converges from above to* μ_1'.

Proof: Observe first that

$$\sigma^\ell = \lambda\mu^\ell + (1 - \lambda)\mu' \tag{4.13}$$

with $0 < \lambda := \frac{1-\psi(\mu')}{\psi(\mu^\ell)-\psi(\mu')} < 1$. Therefore, since ψ is convex by Lemma 4.6.5, we obtain that $\psi(\sigma^\ell) \leq \lambda\psi(\mu^\ell) + (1 - \lambda)\psi(\mu')$. It is now easy to check that $\lambda\psi(\mu^\ell) + (1 - \lambda)\psi(\mu') = 1$ and so $\psi(\sigma^\ell) \leq 1 = \psi(\mu_1')$. Moreover, since $\mu^\ell < \mu'$ it follows that $\sigma^\ell < \mu'$ and since ψ is decreasing in the interval $]-\infty, \mu']$ and $\psi(\sigma^\ell) \leq \psi(\mu_1')$ this yields $\sigma^\ell \geq \mu_1'$. Observe now

$$\begin{aligned}\sigma^{\ell+1} \geq \sigma^\ell \quad &\Leftrightarrow \quad \frac{1 - \psi(\mu')}{\psi(\mu^{\ell+1}) - \psi(\mu')}(\mu^{\ell+1} - \mu') \geq \frac{1 - \psi(\mu')}{\psi(\mu^\ell) - \psi(\mu')}(\mu^\ell - \mu') \\ &\Leftrightarrow \quad \frac{\mu^{\ell+1} - \mu'}{\psi(\mu^{\ell+1}) - \psi(\mu')} \geq \frac{\mu^\ell - \mu'}{\psi(\mu^\ell) - \psi(\mu')}\end{aligned}$$

and this is immediately clear from the convexity of ψ. The remainder of the proof follows easily from the continuity of ψ by computing the limit in (4.12) after observing that $\psi(\mu_1') - \psi(\mu') > 0$. $\square$

Clearly, Lemma 4.6.7 yields the following relation

$$\sigma^\ell - \mu^\ell \leq \varepsilon \Rightarrow 0 \leq \mu_1' - \mu^\ell \leq \varepsilon. \tag{4.14}$$

Obviously, to find μ_2' it is enough to take $\boldsymbol{b} := \boldsymbol{\alpha}_2$, $\boldsymbol{u} := \boldsymbol{\alpha}_1 - \boldsymbol{\alpha}_2$ and then apply exactly the same steps as for computing μ_1'.

As a final remark we note for $1 < p < 2$ that the unit ℓ_2-circle plays the same role as the unit ℓ_∞-circle, i.e. there are two intersection points of the line $\mathcal{K}$ and the unit ℓ_2-circle (let us denote them by $\boldsymbol{\beta}_j$, $j = 1, 2$) and they satisfy

$$[\boldsymbol{a}_1, \boldsymbol{a}_2] \subset [\boldsymbol{\beta}_1, \boldsymbol{\beta}_2] \subset [\boldsymbol{\alpha}_1, \boldsymbol{\alpha}_2].$$

Hence, for $1 < p < 2$ the points $\boldsymbol{\beta}_j$, also analytically computable by (1b), can be used instead of the points $\boldsymbol{\alpha}_j$ and they provide a better first approximation with the same properties.

So we can find both values of μ_j', $j = 1, 2$, and consequently both values of the normal vectors $\boldsymbol{a}_j$ by solving a pair of independent nonlinear equations of the form $||\boldsymbol{b} + \mu \boldsymbol{u}||_p = 1$, with given $\boldsymbol{b}$ and $\boldsymbol{u}$, using the Newton-Raphson method.

Let us assume now that the two values of $\boldsymbol{a}_j$ are known. We still need to find the pair of tangent points $\boldsymbol{t}^L$ and $\boldsymbol{t}^R$ in order to apply Lemma 4.6.4.

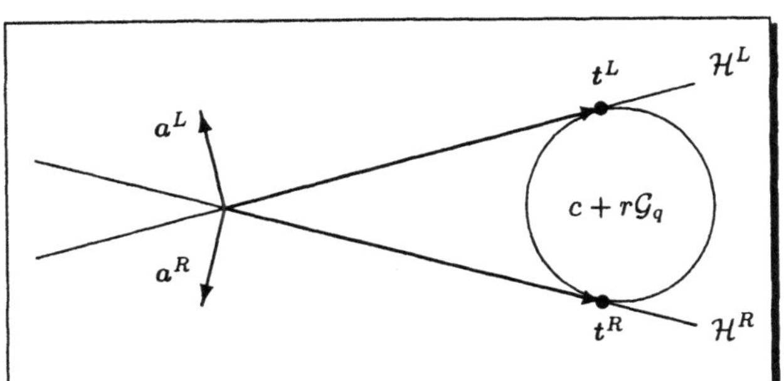

Figure 4.17: Assigning *left* and *right*

Recall that each $\boldsymbol{a}$ is orthogonal to the corresponding $\mathcal{H}$ and is pointing to the half-plane not containing $\boldsymbol{c}+r\mathcal{G}_p$.

Observe that

$$\arg\max\{\boldsymbol{a}_j^\top \boldsymbol{x} : \boldsymbol{x} \in \boldsymbol{c}+r\mathcal{G}_q\}$$

equals

$$\boldsymbol{c}+r\arg\max\{\boldsymbol{a}_j^\top \boldsymbol{y} : \boldsymbol{y} \in \mathcal{G}_q\}.$$

By Hölder's inequality (Ref. [106]) it follows that $\max\{\boldsymbol{a}_j^\top \boldsymbol{y} : \boldsymbol{y} \in \mathcal{G}_q\} = \|\boldsymbol{a}_j\|_p$ with $\frac{1}{p}+\frac{1}{q}=1$. Moreover, the solution of this problem is given by the vector $\boldsymbol{y}^* := (y_1^*, y_2^*)$ with $y_i^* := \operatorname{sign}(a_{j_i})|a_{j_i}|^{p/q}\|\boldsymbol{a}_j\|_p^{1-p}$, $i=1,2$, and $\operatorname{sign}(x)$ denoting the sign function. Hence the point

$$\boldsymbol{t}_j := \arg\max\{\boldsymbol{a}_j^\top \boldsymbol{x} : \boldsymbol{x} \in \boldsymbol{c}+r\mathcal{G}_q\}$$

can be computed for $j=1,2$. Now it is trivial to label them as *left* and *right*, see Figure 4.17.

Corresponding to each approximation μ^ℓ of μ_1' we can derive an approximation $\boldsymbol{\tau}$ of $\boldsymbol{t}$. The question that arises naturally concerns now the "safety" of such an approximation regarding the optimality of the underlying location problem. Observe that the decision on optimality is taken by Algorithm 4.2 when applied to check whether $\mathbf{0}$ belongs to the convex hull of the set of "approximated" points $\{\boldsymbol{\tau}_i^L, \boldsymbol{\tau}_i^R : 1 \le i \le \hat{\kappa}\}$.

Consider $\boldsymbol{a}^\ell := \boldsymbol{b}+\mu^\ell \boldsymbol{u}$. Since $\boldsymbol{a}^\ell$ converges to $\boldsymbol{a}$ from outside the unit ℓ_p-circle, it follows that $\|\boldsymbol{a}^\ell\|_p \ge 1$ and $\boldsymbol{a}^\ell \to \boldsymbol{a}$, i.e. $\|\boldsymbol{a}^\ell\|_p \downarrow 1$. So with respect to system (4.5), $\boldsymbol{a}^\ell$ satisfies the following system

$$\begin{cases} \boldsymbol{c}'^\top \boldsymbol{a}^\ell & = 1 \\ \|\boldsymbol{a}^\ell\|_p & > 1 \end{cases},$$

and so

$$\max\{\boldsymbol{a}^{\ell\top} \boldsymbol{y} : \boldsymbol{y} \in \mathcal{G}_q\} = \|\boldsymbol{a}^\ell\|_p > 1.$$

This implies

$$a^{\ell^T} c + r \max\{a^{\ell^T} y : y \in \mathcal{G}_q\} = -r a^{\ell^T} c' + r\|a^\ell\|_p > 0,$$

and hence there exists some $x \in c + r\mathcal{G}_q$ such that $a^{\ell^T} x > 0$. This yields that the hyperplane $\mathcal{H}^\ell := \{z \in \mathbf{R}^2 : a^{\ell^T} z = 0\}$ is secant to $c + r\mathcal{G}_q$ and τ is a "lower" estimate of t, i.e. it belongs to the cone generated by $\mathcal{H}^L$ and $\mathcal{H}^R$ (see the proof of Lemma 4.6.4).

This guarantees that if the output of Algorithm 4.2 regarding the set of approximations $\{\tau_i^L, \tau_i^R : 1 \le i \le \hat{\kappa}\}$ is *YES* then the true answer is also *YES* and therefore, regarding the underlying location problem no false optimality is detected, making this decision a "safe" one. Of course, a point may be optimal and due to the approximation used be identified as nonoptimal. This is in general safer than the opposite situation.

However, a "safe" *NO* can be produced, by the same reasoning, if the roles of μ^ℓ and σ^ℓ are reversed. This time *YES* would be "unsafe" but in early stages of the underlying optimization procedure it may be interesting to first expect a sequence of negative answers and only after the first *YES* is reported switch to the safe *YES* form.

It only remains to analyze the complexity of solving (D'). Since this makes use of an approximation computed by means of the Newton-Raphson method, an accuracy has to be prescribed.

Theorem 4.6.1. *If the Newton-Raphson method is applied with starting point* 0 *to solve the equation* $\psi(\mu) - 1 = 0$ *it follows that an accuracy of* $\varepsilon > 0$ *is met in* $\mathcal{O}(\log_{0.82}(\varepsilon))$ *iterations.*

Proof: Recall (4.6). It follows that the iterates are generated as $\mu^{\ell+1} := \phi(\mu^\ell)$ with $\phi(\mu) := \mu + \frac{1-\psi(\mu)}{\psi'(\mu)}$. From this point on, unless stated otherwise, μ is assumed to belong to the interval $[0, \mu_1'[$. Recall from Lemma 4.6.6 that the sequence of iterates generated by the Newton-Raphson method belongs to this interval. Computing the derivative of this function yields

$$\phi'(\mu) = \frac{\overbrace{\psi''(\mu)}^{>0}\overbrace{(1-\psi(\mu))}^{<0}}{\underbrace{\psi'(\mu)^2}_{>0}} < 0$$

and since $\psi'(\mu)$ has been computed in (4.10) we now get

$$\psi''(\mu) = (1-p)\left(\frac{\psi'(\mu)^2}{\psi(\mu)} - \psi(\mu)^{1-p}\sum_{i=1}^{2}|b_i + \mu u_i|^{p-2}u_i^2\right)$$

and finally

$$\phi'(\mu) = \underbrace{(1-p)}_{<0}\left(\underbrace{\frac{\psi(\mu)-1}{\psi(\mu)}}_{>0} - \underbrace{\frac{\psi(\mu)^{1-p}(\psi(\mu)-1)}{\psi'(\mu)^2}\sum_{i=1}^{2}|b_i + \mu u_i|^{p-2}u_i^2}_{>0}\right) < 0.$$

Hence

$$|\phi'(\mu)| = (p-1)\left(\frac{\psi(\mu)-1}{\psi(\mu)} - \frac{\psi(\mu)^{1-p}(\psi(\mu)-1)}{\psi'(\mu)^2}\sum_{i=1}^{2}|b_i + \mu u_i|^{p-2}u_i^2\right).$$

It follows that

$$\begin{aligned} |\phi'(\mu)| &< (p-1)\frac{\psi(\mu)-1}{\psi(\mu)} \\ &< (p-1)\frac{\psi(0)-1}{\psi(\mu_1')} = (p-1)(\psi(0)-1) = (p-1)(\|\boldsymbol{b}\|_p - 1). \end{aligned}$$

Recall now by the construction of $\boldsymbol{b}$ (see Figure 4.15) that either $\|\boldsymbol{b}\|_\infty = 1$ or $\|\boldsymbol{b}\|_2 = 1$. In the later case $\|\boldsymbol{b}\|_\infty < 1$ and hence we have that in any case $\|\boldsymbol{b}\|_\infty \leq 1$. Now using $\boldsymbol{b} \in \mathbf{R}^2$ and since each $|b_i| \leq 1$ it follows that $\|\boldsymbol{b}\|_p \leq 2^{1/p}$. This yields $|\phi'(\mu)| \leq (p-1)(2^{1/p}-1)$. In order to bound $2^{1/p}$ we consider the Taylor expansion (Ref. [92]) around 0 of 2^x until order 2. Notice that $1 < p < \infty$ and this leads to $0 < 1/p < 1$. Hence we have

$$2^x = 1 + \log(2)x + \frac{2^\xi \log(2)^2}{2}x^2$$

with ξ some value in the interval $]0,1[$. Since $2^\xi < 2$ we finally obtain

$$\begin{aligned}(p-1)(2^{1/p}-1) &< (p-1)\left(1+\log(2)\frac{1}{p}+\frac{2\log(2)^2}{2}\left(\frac{1}{p}\right)^2-1\right) \\ &= \log(2)\frac{p-1}{p}+\log(2)^2\frac{p-1}{p^2}.\end{aligned}$$

Observe now that $(p-1)/p < 1$ and consider $(p-1)/p^2$. This function has a maximum in $p = 2$ and hence $(p-1)/p^2 \leq 1/4$. Thus

$$(p-1)(2^{1/p}-1) < \log(2) + \frac{\log(2)^2}{4} < 0.82$$

and we conclude that

$$|\phi'(\mu)| < 0.82$$

holds for $\mu \in [0, \mu_1']$. Recalling the iterative procedure $\mu^{\ell+1} = \phi(\mu^\ell)$ and observing that $\mu_1' = \phi(\mu_1')$, it follows by the mean-value theorem (Ref. [92]) that

$$\mu^{\ell+1} - \mu_1' = \phi(\mu^\ell) - \phi(\mu_1') = \phi'(\xi^{\ell+1})(\mu^\ell - \mu_1')$$

and so

$$|\mu^{\ell+1} - \mu_1'| \leq 0.82|\mu^\ell - \mu_1'|.$$

Iterating this inequality and observing by Theorem 4.6.6 that all the iterates belong to $[0, \mu_1'] \subseteq [0,1]$ we obtain for any $\ell \geq 0$

$$|\mu^\ell - \mu_1'| \leq 0.82^\ell |\mu^0 - \mu_1'| \leq 0.82^\ell.$$

Using the above inequality and since $0.82^\ell < \varepsilon$ holds for $\ell = \lceil \log(\varepsilon)/\log(0.82) \rceil$ the result follows. □

Observe that the derivation of the previous result resorts to rather loose bounds. For instance Theorem 4.6.1 guarantees that within 120 iterations an accuracy of $\varepsilon = 5 \times 10^{-11}$ is achieved, while on the other hand our computational experience reported in the next section does not include any instance for which more than 6 iterations were necessary to guarantee an accuracy of 5×10^{-16}...

However, the previous theorem suffices to guarantee the following result.

Corollary 4.6.1. *If an accuracy $\varepsilon > 0$ is specified to terminate the Newton-Raphson procedure used to convert (D') into (D) then the decision problem (D') is solved with a total effort of $\mathcal{O}(\hat{\kappa} \log_{0.82}(\varepsilon))$ for the corresponding accuracy .*

Proof: Observe first that each iteration of the Newton-Raphson procedure is carried out with an effort only dependent on the space dimension, 2, which can be considered constant. The construction of the set of points $\{\tau_i^L, \tau_i^R : 1 \leq i \leq \hat{\kappa}\}$ is achieved with $2\hat{\kappa}$ independent executions of the Newton-Raphson procedure which by Theorem 4.6.1 requires $\mathcal{O}(\log_{0.82}(\varepsilon))$ iterations. The result follows now from Lemma 4.6.3. □

Observe that in the implementation of this procedure we check the difference $\sigma^\ell - \mu^\ell$ and apply (4.14) instead of waiting until the required number of iterations has been performed. The description of this implementation and the results obtained is the subject of the next subsection.

4.6.3. Computational results

The computational experience was carried out over 360 uncorrelated instances of the problem (D'). Those instances were randomly generated in the following way.

The number $\hat{\kappa}$ of ℓ_p-circles belongs to $\{10, 25, 50, 100, 250, 500\}$. For the ℓ_p-norms used we take $p \in \{1.1, 1.5, 1.9, 2.1, 3.0\}$. Finally, the tolerance parameter used in the stopping rule is given by $\varepsilon := 5 \times 10^{-16}$. This "unrealistic" precision, only possible by means of the Extended Precision of Turbo Pascal (see discussion in Subsection 3.4), was deliberately adopted to test the Newton-Raphson method to the limit.

The procedure to generate the circles can be described as follows. All the centers are uniformly generated within the square $[-10, 90] \times [-10, 90]$. Subsequently we randomly generate one radius for each circle in the interval $]0, 3/4\|\mathbf{c}\|_q[$. Hence, each circle is guaranteed not to include $\mathbf{0}$.

The results obtained are included in Table 4.1. Each line of the table corresponds to averages of 10 uncorrelated examples. The first three columns of Table 4.1 describe the problem characteristics. The last row in each group, mixed norms, is generated by randomly selecting $p_i \in \{1.1, 1.5, 1.9, 2.1, 3.0\}$ for each circle $1 \leq i \leq \hat{\kappa}$.

The following two columns describe the behavior of the *Newton-Raphson* algorithm to find the two tangent hyperplanes, i.e. *it* contains the average number of iterations per execution of the Newton-Raphson algorithm, *max it* contains the maximum number of iterations taken by an execution of the same algorithm and *% time* contains the percentage of the total average computation time spent on finding the hyperplanes.

The next pair of columns describes the behaviour of *Algorithm 4.2*, i.e. *% time* is the percentage of the total average time taken by Algorithm 4.2, and *% yes* is the percentage of problems where $\mathbf{0}$ was found to be in the convex hull of the corresponding tangent points.

Finally, the last column includes the total average execution times in seconds obtained on the Compaq Prolinea. Since in Table 4.1 we generated instances with a high percentage of *YES* answers we also applied the algorithm to the probably more difficult instances with a high percentage of *NO* answers. This is achieved by simply generating the centers within the square $[-1, 99] \times [-1, 99]$. These results are summarized in Table 4.2.

Some interesting conclusions can be drawn from these results. First of all, the number of Newton-Raphson iterations required to compute each hyperplane with the given precision is always very low. Secondly, Algorithm 4.2 proves to be extremely efficient in practice (remember that the number of points of its input is twice the number of circles). Finally, when the percentage of time taken by each stage of the algorithm is considered the distribution of the total effort exhibits a very strong regularity, i.e. around 97% of the time is taken by the conversion of (D) into (D') and the remaining 3% are taken by Algorithm 4.2.

problem		*Newton-Raphson*			*Algorithm 4.2*		*total*
$\hat{\kappa}$	p	*it*	*max it*	*% time*	*% time*	*% yes*	*time*
10	1.1	2.4	5	96.7	3.3	20.0	0.006
10	1.5	3.9	6	98.0	2.0	20.0	0.008
10	1.9	4.1	5	97.9	2.1	30.0	0.008
10	2.1	3.8	4	98.2	1.8	50.0	0.008
10	3.0	4.1	5	98.1	1.9	20.0	0.009
10	mixed	3.7	6	97.9	2.1	40.0	0.008
25	1.1	2.5	5	97.7	2.3	50.0	0.015
25	1.5	3.9	6	98.5	1.5	60.0	0.021
25	1.9	4.2	6	98.4	1.6	50.0	0.021
25	2.1	3.7	4	98.3	1.7	50.0	0.020
25	3.0	4.2	5	98.7	1.3	100.0	0.022
25	mixed	3.6	6	97.9	2.1	40.0	0.020
50	1.1	2.5	6	98.0	2.0	70.0	0.029
50	1.5	3.9	6	98.6	1.4	80.0	0.041
50	1.9	4.1	6	98.9	1.1	100.0	0.042
50	2.1	3.8	4	98.7	1.3	90.0	0.039
50	3.0	4.1	5	98.9	1.1	100.0	0.044
50	mixed	3.7	6	98.7	1.3	100.0	0.039
100	1.1	2.6	6	99.1	0.9	100.0	0.058
100	1.5	3.9	6	99.0	1.0	100.0	0.083
100	1.9	4.1	6	98.8	1.2	100.0	0.084
100	2.1	3.7	4	98.8	1.2	100.0	0.077
100	3.0	4.1	5	99.4	0.6	100.0	0.088
100	mixed	3.7	6	99.2	0.8	100.0	0.078
250	1.1	2.6	6	99.7	0.3	100.0	0.145
250	1.5	3.9	6	99.8	0.2	100.0	0.204
250	1.9	4.1	6	99.6	0.4	100.0	0.208
250	2.1	3.7	4	99.6	0.4	100.0	0.192
250	3.0	4.1	5	99.7	0.3	100.0	0.217
250	mixed	3.7	6	99.7	0.3	100.0	0.194
500	1.1	2.6	6	99.8	0.2	100.0	0.291
500	1.5	3.8	6	99.8	0.2	100.0	0.408
500	1.9	4.1	6	99.8	0.2	100.0	0.414
500	2.1	3.7	4	99.8	0.2	100.0	0.384
500	3.0	4.1	5	99.8	0.2	100.0	0.434
500	mixed	3.7	6	99.8	0.2	100.0	0.387

Table 4.1: Results of the decision algorithm, easy case, $\varepsilon = 5 \times 10^{-16}$

problem		*Newton-Raphson*			*Algorithm 4.2*		*total*
$\hat{\kappa}$	p	*it*	*max it*	*% time*	*% time*	*% yes*	*time*
10	1.1	2.5	6	96.8	3.2	0.0	0.006
10	1.5	3.8	6	97.8	2.2	0.0	0.008
10	1.9	4.2	5	97.8	2.2	0.0	0.009
10	2.1	3.8	4	97.7	2.3	0.0	0.008
10	3.0	4.0	5	97.9	2.1	0.0	0.009
10	mixed	3.7	5	97.7	2.3	0.0	0.008
25	1.1	2.5	6	96.7	3.3	0.0	0.015
25	1.5	3.9	6	97.7	2.3	0.0	0.021
25	1.9	4.1	6	97.7	2.3	0.0	0.021
25	2.1	3.7	4	97.6	2.4	10.0	0.020
25	3.0	4.2	5	97.9	2.1	20.0	0.022
25	mixed	3.6	5	97.5	2.5	0.0	0.020
50	1.1	2.5	6	96.6	3.4	0.0	0.030
50	1.5	3.8	6	97.6	2.4	0.0	0.042
50	1.9	4.1	6	97.6	2.4	10.0	0.042
50	2.1	3.8	4	97.4	2.6	0.0	0.039
50	3.0	4.1	5	97.7	2.3	0.0	0.045
50	mixed	3.7	6	97.4	2.6	0.0	0.040
100	1.1	2.5	6	96.5	3.5	0.0	0.059
100	1.5	3.8	6	97.5	2.5	0.0	0.084
100	1.9	4.1	6	97.5	2.5	0.0	0.085
100	2.1	3.7	4	97.4	2.6	10.0	0.078
100	3.0	4.1	5	98.3	1.7	80.0	0.088
100	mixed	3.7	6	97.5	2.5	10.0	0.079
250	1.1	2.6	6	96.8	3.2	10.0	0.150
250	1.5	3.8	6	97.7	2.3	20.0	0.208
250	1.9	4.1	6	97.6	2.4	20.0	0.212
250	2.1	3.7	4	97.6	2.4	20.0	0.196
250	3.0	4.1	5	98.4	1.6	100.0	0.220
250	mixed	3.7	6	97.6	2.4	30.0	0.197
500	1.1	2.6	6	96.6	3.4	10.0	0.300
500	1.5	3.8	6	97.7	2.3	30.0	0.416
500	1.9	4.1	6	97.7	2.3	20.0	0.423
500	2.1	3.7	4	98.2	1.8	70.0	0.389
500	3.0	4.1	5	98.7	1.3	100.0	0.438
500	mixed	3.7	6	98.4	1.6	90.0	0.392

Table 4.2: Results of the decision algorithm, difficult case, $\varepsilon = 5 \times 10^{-16}$

4.7. Computational results

In order to test Algorithm 3.13 we considered the following problem, where $\boldsymbol{H}_i$ is a positive definite matrix for each $1 \leq i \leq n$.

$$\min_{\boldsymbol{x} \in \mathbb{R}^s} \max_{1 \leq i \leq n} (\boldsymbol{x} - \boldsymbol{d}_i)^\top \boldsymbol{H}_i (\boldsymbol{x} - \boldsymbol{d}_i)$$

The objective function in this problem is the maximum of convex quadratic functions. It turns out that this problem is an instance of (L).

Since $\|\boldsymbol{x}\|_{\boldsymbol{H}^{-1}} := \sqrt{\boldsymbol{x}^\top \boldsymbol{H} \boldsymbol{x}}$ is a norm for every positive definite matrix $\boldsymbol{H}$ this problem corresponds to minimizing $\max_{1 \leq i \leq n} \varphi_i(\gamma(\boldsymbol{x}))$ with

$$\gamma(\boldsymbol{x}) := (\|\boldsymbol{x} - \boldsymbol{d}_1\|_{\boldsymbol{H}_1^{-1}}, \ldots, \|\boldsymbol{x} - \boldsymbol{d}_n\|_{\boldsymbol{H}_n^{-1}})$$

and $\varphi_i : \mathbb{R}^n_+ \longrightarrow \mathbb{R}$ given by $\varphi_i(\boldsymbol{z}) := z_i^2$.

Observe that $f_i(\boldsymbol{x}) := \varphi_i(\gamma(\boldsymbol{x})) = (\boldsymbol{x} - \boldsymbol{d}_i)^\top \boldsymbol{H}_i (\boldsymbol{x} - \boldsymbol{d}_i)$ is differentiable for every $\boldsymbol{x}$ and its gradient $\nabla f_i(\boldsymbol{x})$ equals $2\boldsymbol{H}_i(\boldsymbol{x} - \boldsymbol{d}_i)$.

We now explain how the data were generated. The space dimension s was taken in $\{5, 10\}$. Each demand point $\boldsymbol{d}_i$ has coordinates randomly generated in the interval $[-5, 5]$. In order to generate positive definite matrices $\boldsymbol{H}_i$ we explore their $\boldsymbol{LDL}^\top$ decomposition (see Section 3.2.3) and generate the elements of the diagonal of $\boldsymbol{D}_i$ in the interval $[1, 2]$ and the elements of the lower triangle of $\boldsymbol{L}_i$ in the interval $[-1, 1]$. By Theorem 4 of Ref. [79] the set of optimal solutions of this problem is a subset of the convex hull of the demand points. Since all the demand points are generated in $[-5, 5]^s$ this hyperrectangle was taken as an outer approximation of the referred convex hull. Hence, for every instance the first center is $\boldsymbol{a}_0 := \boldsymbol{0}$ and the first radius $r_0 := r = 5\sqrt{s} + 1$.

In order to generate constrained problems (of the same instances solved as unconstrained) we took as a feasible region the hyperrectangle $[0, 2]^s$.

Finally, the execution is stopped with a relative error, see Section 3.3, of ε equal to 5×10^{-6} or to 5×10^{-11}. No optimality test was applied. Table 4.3 and Table 4.5 list the results obtained for the constrained problems with $\varepsilon = 5 \times 10^{-6}$ and $\varepsilon =$

5×10^{-11} respectively, while Table 4.4 and Table 4.6 list the corresponding results for unconstrained problems.

We describe the columns of the tables describing the results for constrained instances since the other tables include a subset of these.

As mentioned in Section 3.4, each line of the tables corresponds to averages of values observed for 10 identical but uncorrelated instances.

Under *problem*, s identifies the space dimension, n the number of demand points (hence the number of quadratic functions) and m (always equal to $2s$) the number of constraints.

Under *central cuts*, *time* identifies the time in seconds measured in the Compaq Prolinea, *it* the number of iterations of the ellipsoid algorithm, *% o* the percentage of *it* which were objective cuts and *% c* the percentage of *it* which were constraint cuts. No norm cut was observed for any instance.

Under *deep cuts*, *time* identifies the time in seconds measured in the Compaq Prolinea, *it* the number of iterations of the ellipsoid algorithm, *% o* the percentage of *it* which were objective cuts and α_o the average depth of such a cut, d the percentage of o which were deep ($\alpha > 0$) and α_d the average depth (average taken only within the cuts that were deep) of a deep objective cut. Finally *% c* is the percentage of *it* which were constraint cuts (always deep) and α_c the corresponding average depth.

Under *% reduction*, *time* corresponds to the percentage by which the application of deep cuts reduces the total time and *it* has a similar meaning for the iterations.

No norm cut was observed for any instance.

problem			*central cuts*				*deep cuts*								*% reduction*	
s	n	m	*time*	*it*	*% o*	*% c*	*time*	*it*	*% o*	α_o	*% d*	α_d	*% c*	α_c	*time*	*it*
5	5	10	0.454	579.2	42.1	57.9	0.333	420.8	41.3	0.0326	62.9	0.0611	58.7	0.0281	26.5	27.3
5	10	10	0.676	582.7	38.9	61.1	0.521	444.6	38.8	0.0274	54.5	0.0557	61.2	0.0257	23.0	23.7
5	20	10	1.096	565.8	48.6	51.4	0.795	407.8	47.8	0.0366	70.2	0.0532	52.2	0.0286	27.5	27.9
10	5	20	5.002	2094.9	33.8	66.2	3.897	1625.1	31.1	0.0203	78.5	0.0283	68.9	0.0084	22.1	22.4
10	10	20	7.549	2160.5	32.8	67.2	5.918	1689.9	30.2	0.0178	77.2	0.0241	69.8	0.0083	21.6	21.8
10	20	20	11.806	2076.0	46.4	53.6	9.037	1586.7	43.8	0.0200	84.5	0.0238	56.2	0.0082	23.5	23.6

Table 4.3: Results for constrained examples, $\varepsilon = 5 \times 10^{-6}$

problem		*central cuts*		*deep cuts*					*% reduction*	
s	*m*	*time*	*it*	*time*	*it*	α_o	*% d*	α_d	*time*	*it*
5	5	0.301	461.4	0.226	344.3	0.0307	83.0	0.0368	24.8	25.4
5	10	0.497	484.2	0.359	346.6	0.0354	85.3	0.0413	27.8	28.4
5	20	0.872	490.7	0.634	354.1	0.0343	85.2	0.0401	27.3	27.8
10	5	2.767	1456.2	2.095	1094.7	0.0153	91.0	0.0168	24.3	24.8
10	10	5.138	1690.7	3.747	1229.4	0.0168	92.4	0.0182	27.1	27.3
10	20	9.146	1750.3	6.481	1238.1	0.0182	92.9	0.0196	29.1	29.3

Table 4.4: Results for unconstrained examples, $\varepsilon = 5 \times 10^{-6}$

problem			*central cuts*				*deep cuts*								*% reduction*	
s	*n*	*m*	*time*	*it*	*% o*	*% c*	*time*	*it*	*% o*	α_o	*% d*	α_d	*% c*	α_c	*time*	*it*
5	5	10	0.870	1110.6	44.8	55.2	0.679	855.6	43.8	0.0259	57.4	0.0559	56.2	0.0240	22.0	23.0
5	10	10	1.310	1127.9	43.5	56.5	1.006	857.1	42.5	0.0284	58.7	0.0572	57.5	0.0253	23.3	24.0
5	20	10	2.164	1117.7	40.4	59.6	1.674	859.0	40.1	0.0250	61.0	0.0478	59.9	0.0251	22.7	23.1
10	5	20	10.414	4362.7	34.3	65.7	8.340	3478.4	31.7	0.0179	79.0	0.0264	68.3	0.0075	19.9	20.3
10	10	20	15.078	4316.1	35.5	64.5	12.129	3463.9	33.3	0.0171	78.5	0.0220	66.7	0.0075	19.6	19.7
10	20	20	24.033	4225.5	46.5	53.5	18.621	3269.1	43.6	0.0197	85.7	0.0230	56.4	0.0077	22.5	22.6

Table 4.5: Results for constrained examples, $\varepsilon = 5 \times 10^{-11}$

problem		*central cuts*		*deep cuts*					*% reduction*	
s	*m*	*time*	*it*	*time*	*it*	α_o	*% d*	α_d	*time*	*it*
5	5	0.599	920.7	0.435	663.5	0.0345	85.7	0.0402	27.3	27.9
5	10	0.986	961.4	0.725	700.8	0.0334	84.1	0.0392	26.5	27.1
5	20	1.739	978.3	1.290	720.1	0.0322	84.2	0.0378	25.8	26.4
10	5	5.956	3131.9	4.507	2353.2	0.0153	92.3	0.0166	24.3	24.9
10	10	10.171	3347.9	7.459	2447.8	0.0168	92.9	0.0181	26.7	26.9
10	20	19.207	3675.9	13.998	2674.1	0.0169	93.3	0.0181	27.1	27.3

Table 4.6: Results for unconstrained examples, $\varepsilon = 5 \times 10^{-11}$

The speculation made in Subsection 3.2.4 that unconstrained instances are solved faster is supported by these results, since the instances are the same, only differing in the presence or not of constraints, and the same initial ellipsoid was taken in any case.

It is also interesting to observe that by computing the average ratio between the number of iterations in the central cut case and the number $s^2 \log(1/\varepsilon)$ (see Theorem 3.2.4) one obtains approximately 1.8 for the constrained case and 1.4 for the unconstrained case. The regularity of these ratios indicates that the ellipsoid method, just as its one-dimensional counterpart, the bisection method, performs consistently according to its worst case.

Finally we observe that the effect of applying deep cuts is very encouraging.

V Quasiconvex Programming

This chapter starts by introducing in Section 5.1 a so-called *quasiconvex program.* In order to solve such a program by the ellipsoid method it should be computationally possible to construct a hyperplane separating a given point from its corresponding strict lower level set. In spite of the convexity of this set such a "computable" separation is not a trivial matter for a quasiconvex function. Therefore, Section 5.2, based on Ref. [39], is devoted to this topic. This section starts by analyzing the cone of descent directions of a quasiconvex function at a given point and using the properties of this cone the local geometry of the corresponding strict lower level set is studied in Subsection 5.2.1. In the case where the cone of descent directions also contains strict descent directions it turns out that the local information given by this smaller cone completely determines all hyperplanes separating the considered point from its strict lower level set. This observation enables us in Subsection 5.2.2 to show that implementing the separation oracle can be done by solving an optimization problem and so it is in principle possible to construct separating hyperplanes if the cone of strict descent directions is nonempty. Moreover, this optimization problem also determines whether the cone of strict descent directions is empty or not. Since solving the optimization problem may be a difficult computational task we discuss in Section 5.3 instances of quasiconvex functions for which this optimization problem can be replaced by an easier membership problem. Since in the absence of strict descent directions the local information given by the cone of descent directions is not sufficient to construct an implementable separation oracle we explain in Section 5.4 what to do in such a situation and the convergence of the corresponding modified ellipsoid algorithm is proven in Section 5.5. Finally, Section 5.6 adapts the ellipsoid method in order to accommodate an easy localization set to assist the implementation of the strategy proposed in Section 5.4 and, taking advantage of the new localization set, the stopping rules proposed in Chapter 3 are strengthened in Section 5.7.

5.1. Introduction

As Chapter 3 and Chapter 4 show, convexity is a very desirable property if one needs to solve a nonlinear program. However, for many applications the assumption of a convex objective function is too restrictive. For this reason, convexity has been generalized in many ways, giving rise to the appropriately named field of *generalized convexity* (Ref. [7, 93]). This field originally appeared in connection with economical applications, where it is reasonable to assume that the lower level sets of an objective function are convex. This is precisely the definition of a quasiconvex function given in Chapter 2. The literature in generalized convexity largely concerns the derivation of subclasses of quasiconvex functions satisfying some additional desirable property (Ref. [7]). In this chapter we do not attempt to depict all these classes but we concentrate on the more general class of quasiconvex functions. These functions also seem from their own definition to be the largest class of functions to which the ellipsoid method can be applied. In fact, as Chapter 3 argues, the ellipsoid method works if and only if each point can be separated from its strict lower level set by means of a hyperplane, and, as shown in Chapter 2, convex analysis proves that convex sets are those sets for which the existence of these separators is guaranteed.

Our quasiconvex optimization problem is now given by

$$(Q) \quad \min\{f(\boldsymbol{x}) : \boldsymbol{x} \in \mathbf{R}^s,\ g(\boldsymbol{x}) \leq 0\}$$

with $f : \mathbf{R}^s \longrightarrow \mathbf{R}$ a continuous quasiconvex objective function and $g : \mathbf{R}^s \longrightarrow \mathbf{R}$ a convex constraint function. In the remainder we assume that $f : \mathbf{R}^s \longrightarrow \mathbf{R}$ is a quasiconvex function with $\boldsymbol{d} \longmapsto f'_+(\boldsymbol{x}_0; \boldsymbol{d})$ a proper function. The definitions of *proper* functions and *Dini upper derivatives* $f'_+(\boldsymbol{x}_0; \boldsymbol{d})$ are given in Chapter 2. Observe, as mentioned in Chapter 3, that there is no loss of generality in considering one single constraint since the maximum of convex functions is convex. Since the same property holds for quasiconvex functions, problem (Q) also includes min-max problems, see Chapter 6.

In order to solve instances of (Q) by a similar technique as the one used to solve instances of (C), one should be able to compute elements from the normal cone of the lower level sets. However, in the case of quasiconvex functions it is essential to consider the strict lower level set since a nonminimal point may belong to a flat region, see Figure 5.2 ahead.

As this chapter shows, quasiconvex programming is somewhere in between convex programming and global optimization. With convex programming it shares the possibility of, in some cases, separating a point from the convex set of points yielding a lower function value. Its *global* character comes from the existence of "bad" points for which local information does not suffice to decide whether lower valued points exist or not. In order to minimize quasiconvex functions in their general form one cannot avoid having to perform some sort of global search.

At a first glance, one expects this situation to rule out the application of the ellipsoid algorithm, since this algorithm relies upon the possibility of deciding how to separate a point from the points being searched. Therefore, we modify the ellipsoid algorithm in such a way that it keeps track of a *box* or *hyperrectangle* containing a minimal point. This hyperrectangle is in general much smaller than the current ellipsoid and can be constructed without increasing the complexity order of the algorithm. This gives the opportunity, in case the center of the current ellipsoid is such a "bad" point, to apply global search techniques over this "easy" set. This global search provides one of two possible answers. It either certificates that the present best known feasible solution is optimal or it replaces the "bad" point by a "good" point from where the ellipsoid method can proceed. Observe that such a "good" point always exists if the present "bad" point is nonoptimal. Details on this and related issues are described in Section 5.4. However, in order to start our analysis we study in the next section the descent directions of a quasiconvex function at a given point and discuss the implication of this for the corresponding lower level set.

5.2. A separation oracle for quasiconvex functions

In contrast with convex functions, not all quasiconvex functions have directional derivatives. Therefore, an important generalization of directional derivatives to consider in this context is given by the *Dini upper derivative* of f at x_0 in the direction

d. This derivative is defined in Chapter 2 as

$$f'_+(x_0; d) := \limsup_{t \downarrow 0} \frac{f(x_0 + td) - f(x_0)}{t}.$$

The next subsection studies in detail the properties of a proper Dini upper derivative of a quasiconvex function f at x_0 and uses these properties to provide a clear understanding of the local geometry of the corresponding strict lower level set. This analysis, also presented by Frenk, Gromicho and Zhang in Ref. [39], is essential for the construction of an oracle separating x_0 from its strict lower level set.

5.2.1. Descent directions and geometry of lower level sets

This subsection mainly derives similar results as those obtained by Crouzeix in Ref. [21–23]. However, while Crouzeix imposes the condition that $d \longmapsto f'_+(x_0; d)$ is a lower semicontinuous (or closed, notice that in this chapter the term lower semicontinuous is preferably used) and proper quasiconvex function we assume the weaker condition that $d \longmapsto f'_+(x_0; d)$ is a proper evenly quasiconvex function (Definition 2.4.2). Despite this weaker assumption it is possible to derive for the above Dini upper derivative similar results by means of easier proofs. Since the main results in this subsection are actually duality results for quasiconvex functions these simple proofs are possible using a more natural generalization (Ref. [35, 78]) of the well-known biconjugate or Fenchel-Moreau theorem for convex functions (Ref. [52, 89]). It turns out that proper evenly quasiconvex functions originate a more symmetrical representation in the dual space than proper lower semicontinuous quasiconvex functions (Ref. [35, 78]) and using this more suitable representation one can give simpler proofs. Moreover, since the definition of an evenly quasiconvex function already "includes" a separation result for convex sets it is also possible to give a very simple and easy proof for this dual representation of proper evenly quasiconvex functions (Ref. [35]). To start this section we observe by the definition of lim sup that $f'_+(x_0; d)$ always exists, i.e. $-\infty \leq f'_+(x_0; d) \leq +\infty$, for any function $f : \mathbb{R}^s \longrightarrow \overline{\mathbb{R}}$ with $f(x_0)$ finite. Moreover, it is easy to verify that $d \longmapsto f'_+(x_0; d)$ is positively homogeneous. If we know additionally that f is quasiconvex then the next result (Ref. [21]) is easy to prove using Lemma 2.4.1.

Lemma 5.2.1. *Let* $f : \mathbf{R}^s \longrightarrow \overline{\mathbf{R}}$ *be a quasiconvex function with* $f(\boldsymbol{x}_0)$ *finite. Then the function* $\boldsymbol{d} \longmapsto f'_+(\boldsymbol{x}_0; \boldsymbol{d})$ *is positively homogeneous and quasiconvex.*

The remaining part of this subsection is now devoted to improve the previous result. It is shown under the condition that $\boldsymbol{d} \longmapsto f'_+(\boldsymbol{x}_0; \boldsymbol{d})$ is a proper evenly quasiconvex function which assumes positive and negative values that this function is actually the minimum of two "nice" convex functions and so one can apply to this function classical techniques of convex analysis. This permits the construction of a separation oracle.

However, before improving the previous result we observe the following. As already pointed out in the introduction, it is crucial for the ellipsoid method to be able to compute an element of the normal cone of $\mathcal{L}_f^<(f(\boldsymbol{x}_0))$ at $\boldsymbol{x}_0$. Clearly, this presumes that the set $\mathcal{L}_f^<(f(\boldsymbol{x}_0))$ is nonempty. A sufficient condition to ensure this is the nonemptiness of the set $\mathcal{D}_f^<(\boldsymbol{x}_0) := \{\boldsymbol{d} \in \mathbf{R}^s : f'_+(\boldsymbol{x}_0; \boldsymbol{d}) < 0\}$ of *strict descent directions* at $\boldsymbol{x}_0$. Unfortunately, in the case of quasiconvex functions, contrary to convex functions (Ref. [52]), this condition is not necessary as shown by $f(x) = x^3$ at 0.

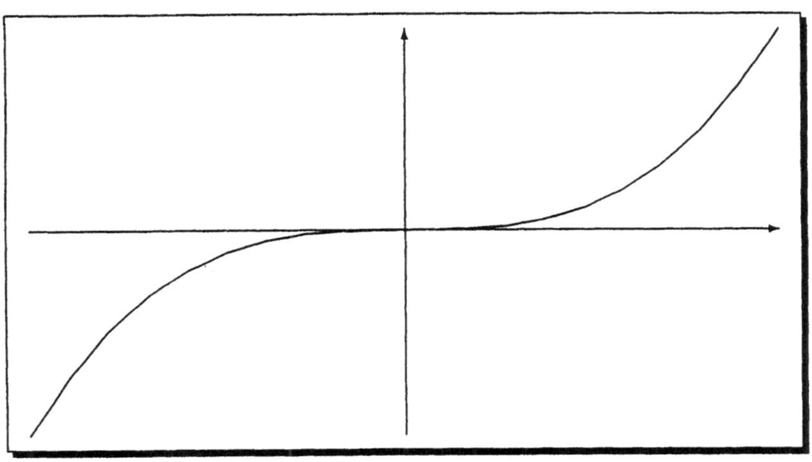

Figure 5.1: A simple but "nasty" quasiconvex function: x^3

This function is differentiable at 0 and its derivative at this point equals 0. Therefore $f'(0; d) = 0$ for every $d \in \mathbf{R}$, while $\mathcal{L}_f^<(0) =]-\infty, 0[$ is clearly nonempty.

For quasiconvex functions a necessary condition is given by the nonemptiness of the set $\mathcal{D}_f^{\leq}(\boldsymbol{x}_0) \setminus \{\boldsymbol{0}\}$ with $\mathcal{D}_f^{\leq}(\boldsymbol{x}_0) := \{\boldsymbol{d} \in \mathbf{R}^s : f'_+(\boldsymbol{x}_0; \boldsymbol{d}) \leq 0\}$ the set of *descent directions*. It turns out that the function $\boldsymbol{d} \longmapsto f'_+(\boldsymbol{x}_0; \boldsymbol{d})$ completely characterizes the normal cone of $\mathcal{L}_f^{<}(f(\boldsymbol{x}_0))$ at $\boldsymbol{x}_0$ if $\mathcal{D}_f^{<}(\boldsymbol{x}_0)$ is nonempty. For $\mathcal{D}_f^{<}(\boldsymbol{x}_0)$ empty we also need global information to find out whether $\mathcal{L}_f^{<}(f(\boldsymbol{x}_0))$ is nonempty or not and so the local information given by $\boldsymbol{d} \longmapsto f'_+(\boldsymbol{x}_0; \boldsymbol{d})$ is insufficient even to decide whether $\boldsymbol{x}_0$ minimizes f or not. We introduce now the set

$$\Gamma_f := \{\boldsymbol{x} \in \mathbf{R}^s : f'_+(\boldsymbol{x}; \boldsymbol{d}) \geq 0 \text{ for every } \boldsymbol{d} \in \mathbf{R}^s\} \tag{5.1}$$

which is sometimes called the set of stationary points. For reasons to be soon clarified we call this the set of *"bad" points.* Before deriving the announced characterization, we observe by Theorem 2.2.9 that the normal cone

$$\mathcal{N}_{\mathcal{L}_f^{<}}(\boldsymbol{x}_0) := \left\{\boldsymbol{x}^\star \in \mathbf{R}^s : \boldsymbol{x}^{\star\mathsf{T}}(\boldsymbol{x} - \boldsymbol{x}_0) \leq 0 \text{ for every } \boldsymbol{x} \in \mathcal{L}_f^{<}(f(\boldsymbol{x}_0))\right\}$$

of $\mathcal{L}_f^{<}(f(\boldsymbol{x}_0))$ at $\boldsymbol{x}_0$ is a proper nonempty convex cone of $\mathbf{R}^s$ if $\mathcal{L}_f^{<}(f(\boldsymbol{x}_0))$ is nonempty. In the next result a partial description of $\mathcal{N}_{\mathcal{L}_f^{<}}(\boldsymbol{x}_0)$ is given by means of the function $\boldsymbol{d} \longmapsto f'_+(\boldsymbol{x}_0; \boldsymbol{d})$ if $\mathcal{D}_f^{<}(\boldsymbol{x}_0)$ is empty. Although trivial this result seems to be new.

Lemma 5.2.2. *For $f : \mathbf{R}^s \longrightarrow \overline{\mathbf{R}}$ a quasiconvex function satisfying $f(\boldsymbol{x}_0)$ finite, $\mathcal{L}_f^{<}(f(\boldsymbol{x}_0))$ nonempty and $\mathcal{D}_f^{<}(\boldsymbol{x}_0)$ empty, i.e. $\boldsymbol{x}_0 \in \Gamma_f$, it follows that $(\mathcal{D}_f^{=}(\boldsymbol{x}_0))^\circ \subseteq \mathcal{N}_{\mathcal{L}_f^{<}}(\boldsymbol{x}_0)$ with $(\mathcal{D}_f^{=}(\boldsymbol{x}_0))^\circ$ denoting the polar of the nonempty convex cone $\mathcal{D}_f^{=}(\boldsymbol{x}_0) := \{\boldsymbol{d} \in \mathbf{R}^s : f'_+(\boldsymbol{x}_0; \boldsymbol{d}) = 0\}$.*

Proof: Since $\mathcal{D}_f^{<}(\boldsymbol{x}_0)$ is empty and f is quasiconvex it must follow by Lemma 5.2.1 that $\mathcal{D}_f^{=}(\boldsymbol{x}_0)$ is a nonempty convex cone. Moreover, the nonemptyness of $\mathcal{L}_f^{<}(f(\boldsymbol{x}_0))$ and the emptiness of $\mathcal{D}_f^{<}(\boldsymbol{x}_0)$ enable us to verify that $\mathcal{L}_f^{<}(f(\boldsymbol{x}_0)) - \boldsymbol{x}_0 \subseteq \mathcal{D}_f^{=}(\boldsymbol{x}_0)$ and so $\text{cone}(\mathcal{L}_f^{<}(f(\boldsymbol{x}_0)) - \boldsymbol{x}_0) \subseteq \mathcal{D}_f^{=}(\boldsymbol{x}_0)$. This implies by (2.2) that $(\mathcal{D}_f^{=}(\boldsymbol{x}_0))^\circ \subseteq (\text{cone}(\mathcal{L}_f^{<}(f(\boldsymbol{x}_0)) - \boldsymbol{x}_0))^\circ = \mathcal{N}_{\mathcal{L}_f^{<}}(\boldsymbol{x}_0)$ and hence the desired result is proven. □

Clearly, the above result reduces to a trivial observation if $\mathcal{D}_f^{=}(\boldsymbol{x}_0)$ equals $\mathbf{R}^s$, as in the case of $f(x) = x^3$ for $x_0 = 0$. In this case, $(\mathcal{D}_f^{=}(\boldsymbol{x}_0))^\circ = \{\boldsymbol{0}\}$ and for $\mathcal{D}_f^{<}(\boldsymbol{x}_0)$

empty a sufficient condition for this to happen is given by $\boldsymbol{x}_0 \in \mathrm{int}(\mathcal{L}_f^{\leq}(f(\boldsymbol{x}_0)))$. The same example, $f(x) = x^3$, shows that this condition is not necessary.

Figure 5.2 provides an interpretation of Lemma 5.2.2. The first picture is drawn in the domain and shows two lower level sets. The one with a dashed boundary is $\mathcal{L}_f^{<}(f(\boldsymbol{b}))$ and the one with a full boundary is $\mathcal{L}_f^{\leq}(f(\boldsymbol{a})) = \mathcal{L}_f^{\leq}(f(\boldsymbol{b}))$. The second picture is drawn in the epigraph space and corresponds to slicing the graph of the function along the line going through $\boldsymbol{a}$ and $\boldsymbol{b}$. Observe first that if $\boldsymbol{x}_0 \in]\boldsymbol{a}, \boldsymbol{b}[$ then $\mathcal{D}_f^{=}(\boldsymbol{x}_0) = \mathbf{R}^s$ and so no useful information is provided. On the other hand, if $\boldsymbol{x}_0 = \boldsymbol{a}$ then $\mathcal{D}_f^{=}(\boldsymbol{x}_0) \neq \mathbf{R}^s$ and so $\left(\mathcal{D}_f^{=}(\boldsymbol{x}_0)\right)^{\circ}$ also contains nonzero elements.

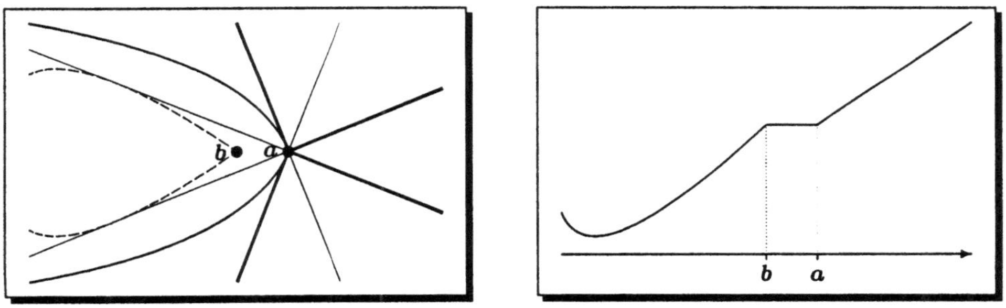

Figure 5.2: Interpretation of the partial description

To discuss the case with $\mathcal{D}_f^{<}(\boldsymbol{x}_0)$ nonempty we first observe that $\mathcal{D}_f^{<}(\boldsymbol{x}_0)$ is a convex subset of the nonempty convex cone $\mathrm{cone}(\mathcal{L}_f^{<}(f(\boldsymbol{x}_0)) - \boldsymbol{x}_0)$. For $\mathcal{D}_f^{<}(\boldsymbol{x}_0)$ nonempty it is shown in the next result that $\mathrm{ri}(\mathrm{cone}(\mathcal{L}_f^{<}(f(\boldsymbol{x}_0)) - \boldsymbol{x}_0))$ equals $\mathrm{ri}(\mathcal{D}_f^{<}(\boldsymbol{x}_0))$. The same result is proven by Crouzeix (Ref. [22]) and for completeness we list a more detailed proof.

Lemma 5.2.3. *If* $f : \mathbf{R}^s \longrightarrow \overline{\mathbf{R}}$ *is a quasiconvex function with* $f(\boldsymbol{x}_0)$ *finite and* $\mathcal{D}_f^{<}(\boldsymbol{x}_0)$ *nonempty then* $\mathrm{ri}(\mathrm{cone}(\mathcal{L}_f^{<}(f(\boldsymbol{x}_0)) - \boldsymbol{x}_0))$ *equals* $\mathrm{ri}(\mathcal{D}_f^{<}(\boldsymbol{x}_0))$.

Proof: Since $\mathcal{D}_f^{<}(\boldsymbol{x}_0) \subseteq \mathrm{cone}(\mathcal{L}_f^{<}(f(\boldsymbol{x}_0)) - \boldsymbol{x}_0)$ and $\mathcal{D}_f^{<}(\boldsymbol{x}_0)$ is nonempty it is sufficient by Theorem 2.2.7 to verify that $\mathrm{ri}(\mathrm{cone}(\mathcal{L}_f^{<}(f(\boldsymbol{x}_0)) - \boldsymbol{x}_0)) \subseteq \mathcal{D}_f^{<}(\boldsymbol{x}_0)$. Consider now some $\boldsymbol{d}_0 \in \mathrm{ri}(\mathrm{cone}(\mathcal{L}_f^{<}(f(\boldsymbol{x}_0)) - \boldsymbol{x}_0))$ and let $\boldsymbol{d} \in \mathcal{D}_f^{<}(\boldsymbol{x}_0) \subseteq \mathrm{cone}(\mathcal{L}_f^{<}(f(\boldsymbol{x}_0)) - \boldsymbol{x}_0)$. By Theorem 2.2.5 there exists some $\mu < 0$ such that $\boldsymbol{d}_\mu := \boldsymbol{d}_0 + \mu(\boldsymbol{d} - \boldsymbol{d}_0) \in \mathrm{ri}(\mathrm{cone}(\mathcal{L}_f^{<}(f(\boldsymbol{x}_0)) - \boldsymbol{x}_0))$ and so $\boldsymbol{d}_0 = \frac{1}{1-\mu}\boldsymbol{d}_\mu - \frac{\mu}{1-\mu}\boldsymbol{d}$. Moreover, since

$d_\mu \in \operatorname{ri}(\operatorname{cone}(\mathcal{L}_f^<(f(x_0))-x_0))$ and by Theorem 2.2.8 it follows that $\mathbf{0}$ does not belong to $\operatorname{ri}(\operatorname{cone}(\mathcal{L}_f^<(f(x_0))-x_0))$ we can find some $t_0>0$ satisfying $f(x_0+t_0d_\mu)<f(x_0)$. Construct now for each $t>0$ the line $\mathcal{L}_t$ going through $x_0+t_0d_\mu$ and $m_t:=x_0+td_0$ and crossing $x_0+\alpha d$ in $n_t:=x_0+\xi_t d$, see Figure 5.3. By the quasiconvexity of f it follows that

$$f(m_t)=f(x_0+td_0)\leq \max\{f(x_0+t_0d_\mu),f(x_0+\xi_t d)\}. \tag{5.2}$$

To compute ξ_t we intersect the line $\mathcal{L}_t$ with the line $x_0+\alpha d$. After some computations we obtain $\xi_t=\frac{-\mu t t_0}{(1-\mu)t_0-t}$. Substituting this into (5.2) yields

$$\begin{aligned}&\frac{f(x_0+td_0)-f(x_0)}{t}\\ &\leq \max\left\{\frac{f(x_0+t_0d_\mu)-f(x_0)}{t},\frac{f\left(x_0+\frac{-\mu t t_0}{(1-\mu)t_0-t}d\right)-f(x_0)}{t}\right\}\end{aligned}$$

and due to $x_0+t_0d_\mu\in\mathcal{L}_f^<(f(x_0))$ and $\lim_{t\downarrow 0}\frac{\xi_t}{t}=\frac{-\mu}{1-\mu}>0$ we obtain

$$f'_+(x_0;d_0)\leq\frac{-\mu}{1-\mu}f'_+(x_0;d)<0.$$

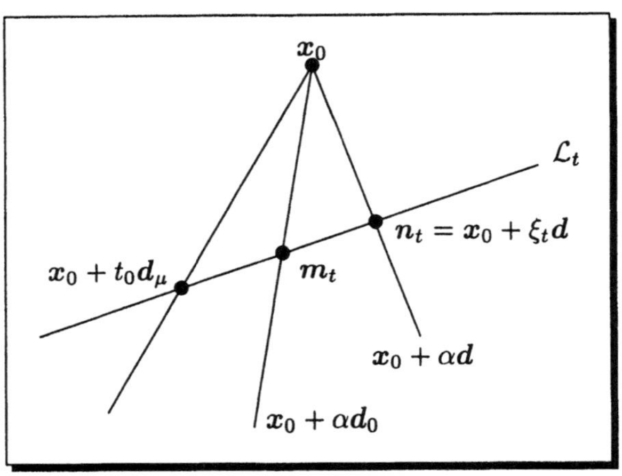

Figure 5.3: Construction of $\mathcal{L}_t$

Hence, $d_0\in\mathcal{D}_f^<(x_0)$ and the desired result is proven. □

By applying Lemma 5.2.3 and Theorem 2.2.6 it follows for $\mathcal{D}_f^{<}(\boldsymbol{x}_0)$ nonempty that $\text{cl}(\mathcal{D}_f^{<}(\boldsymbol{x}_0)) = \text{cl}(\text{cone}(\mathcal{L}_f^{<}(f(\boldsymbol{x}_0)) - \boldsymbol{x}_0))$. Similar as for convex functions (Chapter 2) this yields

$$\begin{aligned}\mathcal{N}_{\mathcal{L}_f^{<}}(\boldsymbol{x}_0) &= \left\{\boldsymbol{x}^\star \in \mathbf{R}^s : \boldsymbol{x}^{\star\top}(\boldsymbol{x} - \boldsymbol{x}_0) \le 0 \text{ for every } \boldsymbol{x} \in \mathcal{L}_f^{<}(f(\boldsymbol{x}_0))\right\} \qquad (5.3)\\ &= (\text{cone}(\mathcal{L}_f^{<}(f(\boldsymbol{x}_0)) - \boldsymbol{x}_0))^\circ = (\text{cl}(\text{cone}(\mathcal{L}_f^{<}(f(\boldsymbol{x}_0)) - \boldsymbol{x}_0)))^\circ\\ &= (\text{cl}(\mathcal{D}_f^{<}(\boldsymbol{x}_0)))^\circ\end{aligned}$$

with $\mathcal{K}^\circ$ denoting the polar cone of $\mathcal{K}$.

Hence, to give an alternative description of the set $\mathcal{N}_{\mathcal{L}_f^{<}}(\boldsymbol{x}_0)$, it is sufficient by (5.3) or Lemma 5.2.2 to show that $\text{cl}(\mathcal{D}_f^{<}(\boldsymbol{x}_0))$ or $\text{cl}(\mathcal{D}_f^{=}(\boldsymbol{x}_0))$ is the polar cone of some other cone. To achieve this we restrict ourselves to so-called *evenly quasiconvex functions*. Such functions, defined in Chapter 2, are those functions having *evenly convex* lower level sets. Observe that every lower semicontinuous quasiconvex function is evenly quasiconvex, since it has closed convex (hence evenly convex) lower level sets. Consider now a quasiconvex function f with $f(\boldsymbol{x}_0)$ finite, $\mathcal{D}_f^{<}(\boldsymbol{x}_0)$ nonempty and $\boldsymbol{d} \longmapsto f'_+(\boldsymbol{x}_0; \boldsymbol{d})$ proper evenly quasiconvex[1] and introduce the function

$$h_-(\boldsymbol{d}) := \begin{cases} f'_+(\boldsymbol{x}_0; \boldsymbol{d}) & \text{if } \boldsymbol{d} \in \text{cl}(\mathcal{D}_f^{<}(\boldsymbol{x}_0)) \\ +\infty & \text{otherwise} \end{cases}.$$

Since $\mathcal{L}_{h_-}^{\le}(\alpha) = \text{cl}(\mathcal{D}_f^{<}(\boldsymbol{x}_0)) \cap \{\boldsymbol{d} \in \mathbf{R}^s : f'_+(\boldsymbol{x}_0; \boldsymbol{d}) \le \alpha\}$ for every $\alpha \in \mathbf{R}$ and by the observations after Definition 2.2.3 the set $\text{cl}(\mathcal{D}_f^{<}(\boldsymbol{x}_0))$ is evenly convex it follows that $\boldsymbol{d} \longmapsto h_-(\boldsymbol{d})$ is also evenly quasiconvex, positively homogeneous and proper. This implies by Theorem 3.5 of Ref. [78], or Lemma 2.7 of Ref. [35], that for every $\boldsymbol{d} \in \mathbf{R}^s$

$$-\infty < h_-(\boldsymbol{d}) = \sup\{k_-(\boldsymbol{x}^\star, \boldsymbol{x}^{\star\top}\boldsymbol{d}) : \boldsymbol{x}^\star \in \mathbf{R}^s\} \qquad (5.4)$$

with

$$k_-(\boldsymbol{x}^\star, r) := \inf\{h_-(\boldsymbol{d}) : \boldsymbol{x}^{\star\top}\boldsymbol{d} \ge r,\ \boldsymbol{d} \in \mathbf{R}^s\}. \qquad (5.5)$$

[1] Crouzeix in Ref. [23] imposes stronger assumptions. It is there assumed that the directional derivative $\boldsymbol{d} \longmapsto f'(\boldsymbol{x}_0; \boldsymbol{d})$ exists and is lower semicontinuous.

In order to study the function $\boldsymbol{d} \longmapsto h_-(\boldsymbol{d})$ we analyze the function $r \longmapsto k_-(\boldsymbol{x}^\star, r)$ for every $\boldsymbol{x}^\star \in \mathbb{R}^s$. As already observed in the beginning of this subsection, relations (5.4) and (5.5) are duality representations for proper evenly quasiconvex functions and these representations are more symmetric than the one used by Crouzeix for proper lower semicontinuous quasiconvex functions (Ref. [23]).

The following result is an immediate consequence of the positive homogeneity of the function $h_- : \mathbb{R}^s \longrightarrow \overline{\mathbb{R}}$.

Lemma 5.2.4. *For each $\boldsymbol{x}^\star \in \mathbb{R}^s$ the function $r \longmapsto k_-(\boldsymbol{x}^\star, r)$ is nondecreasing and positively homogeneous.*

Proof: Obvious. □

By the above lemma and the symmetric relations (5.4) and (5.5) it is possible to derive stronger results for $r \longmapsto k_-(\boldsymbol{x}^\star, r)$.

Lemma 5.2.5. *For $\mathcal{D}_f^<(\boldsymbol{x}_0)$ nonempty it follows that:*

1. *If $\boldsymbol{x}^\star$ does not belong to $(\mathrm{cl}(\mathcal{D}_f^<(\boldsymbol{x}_0)))^\circ$ then $k_-(\boldsymbol{x}^\star, r) = -\infty$ for every $r \in \mathbb{R}$.*
2. *If $\boldsymbol{x}^\star$ belongs to $(\mathrm{cl}(\mathcal{D}_f^<(\boldsymbol{x}_0)))^\circ$ and there exists some $\boldsymbol{d}_0 \in \mathcal{D}_f^<(\boldsymbol{x}_0)$ with $\boldsymbol{x}^{\star\mathsf{T}}\boldsymbol{d}_0 = 0$ then*
$$k_-(\boldsymbol{x}^\star, r) = \begin{cases} +\infty & \text{for } r > 0 \\ -\infty & \text{for } r \leq 0 \end{cases}.$$
3. *If $\boldsymbol{x}^\star$ belongs to $\mathcal{S}$ with $\mathcal{S} := \{\boldsymbol{x}^\star \in \mathbb{R}^s : \boldsymbol{x}^{\star\mathsf{T}}\boldsymbol{d} < 0 \text{ for every } \boldsymbol{d} \in \mathcal{D}_f^<(\boldsymbol{x}_0)\}$ then*
$$k_-(\boldsymbol{x}^\star, r) = \begin{cases} +\infty & \text{for } r > 0 \\ 0 & \text{for } r = 0 \\ -rk_-(\boldsymbol{x}^\star, -1) & \text{for } r < 0 \end{cases}$$
and $-\infty \leq k_-(\boldsymbol{x}^\star, -1) < 0$.

Proof: If $x^\star$ does not belong to $(\mathrm{cl}(\mathcal{D}_f^<(x_0)))^\circ$ one can find some $d_0 \in \mathcal{D}_f^<(x_0)$ with $x^{\star\top}d_0 > 0$. Hence, for every $r > 0$ we obtain by Lemma 5.2.4 and (5.5) that

$$\begin{aligned} k_-(x^\star, r) &= (x^{\star\top}d_0)^{-1}k_-(x^\star, rx^{\star\top}d_0) \\ &\le (x^{\star\top}d_0)^{-1}h_-(rd_0) = r(x^{\star\top}d_0)^{-1}h_-(d_0) < 0 \end{aligned}$$

and this yields $\lim_{r\uparrow\infty} k_-(x^\star, r) = -\infty$. Since $r \longmapsto k_-(x^\star, r)$ is nondecreasing the result in 1. follows. To verify the result in 2. we observe for $r > 0$ and $x^\star \in (\mathrm{cl}(\mathcal{D}_f^<(x_0)))^\circ$ that $x^{\star\top}d \ge r > 0$ implies that d belongs to the complement $(\mathrm{cl}(\mathcal{D}_f^<(x_0)))^c$ of the set $\mathrm{cl}(\mathcal{D}_f^<(x_0))$. Hence, $h_-(d) = +\infty$ for every $d \in \mathbb{R}^s$ satisfying $x^{\star\top}d \ge r > 0$ and by (5.5) this implies $k_-(x^\star, r) = +\infty$. Furthermore, since there exists some $d_0 \in \mathcal{D}_f^<(x_0)$ with $x^{\star\top}d_0 = 0$ it follows by (5.5) for every $\alpha > 0$ that $k_-(x^\star, 0) \le h_-(\alpha d_0) = \alpha h_-(d_0) < 0$ and so $k_-(x^\star, 0) = -\infty$. Due to the fact that $r \longmapsto k_-(x^\star, r)$ is nondecreasing we obtain $k_-(x^\star, r) = -\infty$ for every $r \le 0$. Finally, if $x^\star$ does not satisfy the conditions of 1. or 2. it must follow that $x^\star \in \mathcal{S}$. Since $\mathbf{0} \in \mathrm{cl}(\mathcal{D}_f^<(x_0))$ and $f'_+(x; \mathbf{0}) = 0$ we obtain $k_-(x^\star, 0) \le h_-(\mathbf{0}) = 0$. Moreover, for every $d \in \mathbb{R}^s$ satisfying $x^{\star\top}d \ge 0$ we have for $x^\star \in \mathcal{S}$ that $d \in \mathcal{D}_f^<(x_0)^c$ and so $h_-(d) \ge 0$. Hence, by (5.5) it follows that $k_-(x^\star, 0) \ge 0$ and by our previous inequality this implies that $k_-(x^\star, 0) = 0$. Finally, by Lemma 5.2.4 we obtain for every $r < 0$ that $k_-(x^\star, r) = -rk_-(x^\star, -1)$. Clearly, since $\mathcal{D}_f^<(x_0)$ is nonempty we have for every $x^\star \in \mathcal{S}$ that $-\infty \le k_-(x^\star, -1) < 0$. □

From the above lemma we obtain the next, which improves a result in Ref. [21, 23].

Lemma 5.2.6. *If the set $\mathcal{D}_f^<(x_0)$ is nonempty and $d \longmapsto f'_+(x_0; d)$ is proper and evenly quasiconvex then $d \longmapsto h_-(d)$ is a lower semicontinuous sublinear function.*

Proof: By (5.4), Lemma 5.2.5 and $d \longmapsto h_-(d)$ proper we obtain that $\mathcal{S}$ is nonempty with $\mathcal{S}$ defined as in 3. of Lemma 5.2.5 and

$$-\infty < h_-(d) = \sup\{k_-(x^\star, x^{\star\top}d) : x^\star \in \mathcal{S}\}. \tag{5.6}$$

Suppose now for every $\boldsymbol{x}^\star \in \mathcal{S}$ that $k_-(\boldsymbol{x}^\star, -1) = -\infty$. This implies by Lemma 5.2.5 that

$$k_-(\boldsymbol{x}^\star, r) = \begin{cases} +\infty & \text{if } r > 0 \\ 0 & \text{if } r = 0 \\ -\infty & \text{if } r < 0 \end{cases}$$

and so by (5.6) it follows that $h_-(\boldsymbol{d}) = -\infty$ for every $\boldsymbol{d} \in \mathcal{D}_f^<(\boldsymbol{x}_0)$. This contradicts h_- being proper and so it must follow that the set $\mathcal{S}_1 := \{\boldsymbol{x}^\star \in \mathcal{S} : -\infty < k_-(\boldsymbol{x}^\star, -1) < 0\}$ is nonempty. By this observation it is clear from (5.6) that

$$-\infty < h_-(\boldsymbol{d}) = \sup\{k_-(\boldsymbol{x}^\star, \boldsymbol{x}^{\star\intercal}\boldsymbol{d}) : \boldsymbol{x}^\star \in \mathcal{S}_1\} \tag{5.7}$$

and since by 3. of Lemma 5.2.5 the function $r \longmapsto k_-(\boldsymbol{x}^\star, r)$, and hence $\boldsymbol{d} \longmapsto k_-(\boldsymbol{x}^\star, \boldsymbol{x}^{\star\intercal}\boldsymbol{d})$, is lower semicontinuous, and sublinear for every $\boldsymbol{x}^\star \in \mathcal{S}_1$, the desired result is obtained by (5.7). □

Finally we can prove the following result. Observe that a similar result is proven in Ref. [21, 23] under the stronger assumption that $\boldsymbol{d} \longmapsto f'_+(\boldsymbol{x}_0; \boldsymbol{d})$ is a proper lower semicontinuous quasiconvex function.

Lemma 5.2.7. *If $\boldsymbol{d} \longmapsto f'_+(\boldsymbol{x}_0; \boldsymbol{d})$ is proper and evenly quasiconvex, $\mathcal{D}_f^<(\boldsymbol{x}_0)$ is nonempty and the function $h_- : \mathbf{R}^s \longrightarrow \overline{\mathbf{R}}$ is given by*

$$h_-(\boldsymbol{d}) = \begin{cases} f'_+(\boldsymbol{x}_0; \boldsymbol{d}) & \textit{if } \boldsymbol{d} \in \operatorname{cl}(\mathcal{D}_f^<(\boldsymbol{x}_0)) \\ +\infty & \textit{otherwise} \end{cases}$$

then $\boldsymbol{d} \longmapsto h_-(\boldsymbol{d})$ is a lower semicontinuous sublinear function with its subgradient set $\partial h_-(\boldsymbol{0})$ at $\boldsymbol{0}$ nonempty. Moreover, the normal cone $\mathcal{N}_{\mathcal{L}_f^<}(\boldsymbol{x}_0)$ of $\mathcal{L}_f^<(f(\boldsymbol{x}_0))$ at $\boldsymbol{x}_0$ equals $\operatorname{cl}(\operatorname{cone}(\partial h_-(\boldsymbol{0})))$.

Proof: By Lemma 5.2.6 the first part is already proven. Since the function h_- is lower semicontinuous and sublinear it follows by Theorem 2.3.5 that h_- is the support function of the closed nonempty convex set $\mathcal{C}_- := \{\boldsymbol{x}^\star \in \mathbf{R}^s : \boldsymbol{x}^{\star\intercal}\boldsymbol{d} \leq h_-(\boldsymbol{d}) \text{ for every } \boldsymbol{d} \in \mathbf{R}^s\}$. Since $\boldsymbol{0} \in \operatorname{cl}(\mathcal{D}_f^<(\boldsymbol{x}_0))$ this implies that $h_-(\boldsymbol{0}) =$

$f'_+(\boldsymbol{x}_0;\mathbf{0}) = 0$ and so $\mathcal{C}_- = \partial h_-(\mathbf{0})$. By these observations and Lemma 2.3.1 we obtain that

$$\begin{aligned}
\mathrm{cl}(\mathcal{D}_f^<(\boldsymbol{x}_0)) &= \{\boldsymbol{d} \in \mathbb{R}^s : h_-(\boldsymbol{d}) \le 0\} \\
&= \{\boldsymbol{d} \in \mathbb{R}^s : \boldsymbol{x}^{\star\intercal}\boldsymbol{d} \le 0 \text{ for every } \boldsymbol{x}^\star \in \partial h_-(\mathbf{0})\} \\
&= (\mathrm{cone}(\partial h_-(\mathbf{0})))^\circ
\end{aligned}$$

and so by (5.3) and Lemma 2.2.2 it finally follows that $\mathcal{N}_{\mathcal{L}_f^<}(\boldsymbol{x}_0) = (\mathrm{cl}(\mathcal{D}_f^<(\boldsymbol{x}_0)))^\circ = (\mathrm{cone}(\partial h_-(\mathbf{0})))^{\circ\circ} = \mathrm{cl}(\mathrm{cone}(\partial h_-(\mathbf{0})))$. □

An interpretation of Lemma 5.2.7 is provided by Figure 5.4. Observe that the similarity between these pictures and those in Figure 2.7 is intentional, since this result, despite the nonconvexity of f and its Dini upper derivative at $\boldsymbol{x}_0$, provides a similar relation between $\mathcal{L}_f^<(f(\boldsymbol{x}_0)) - \boldsymbol{x}_0$ and $\partial h_-(\mathbf{0})$ as between $\mathcal{L}_\varphi^{\le}(\varphi(\boldsymbol{x}_0)) - \boldsymbol{x}_0$ and $\partial\varphi(\boldsymbol{x}_0)$ for a convex function φ.

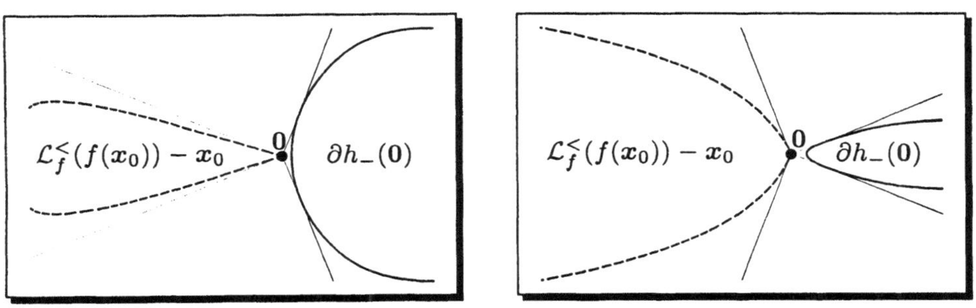

Figure 5.4: The normal cone to the strict lower level set in the favorable case

Compare now this figure with Figure 5.2, The situation described in Figure 5.4 corresponds to taking $\boldsymbol{x}_0 = \boldsymbol{b}$ in Figure 5.2.

Since h_- is sublinear it follows that $\partial h_-(\boldsymbol{d}) \subseteq \partial h_-(\mathbf{0}) \subseteq \mathcal{N}_{\mathcal{L}_f^<}(\boldsymbol{x}_0)$ for every $\boldsymbol{d} \in \mathrm{dom}(h_-) = \mathrm{cl}(\mathcal{D}_f^<(\boldsymbol{x}_0))$. If additionally, $\mathcal{D}_f^<(\boldsymbol{x}_0)$ is a convex cone of dimension s this implies by Theorem 2.3.4 that h_- is differentiable on a dense subset of $\mathrm{int}(\mathcal{D}_f^<(\boldsymbol{x}_0))$, and so we can conclude for a point belonging to this dense subset that $\nabla h_-(\boldsymbol{d}) \in \mathcal{N}_{\mathcal{L}_f^<}(\boldsymbol{x}_0)$.

As in Ref. [23] we may also introduce the function $h_+ : \mathbf{R}^s \longrightarrow \overline{\mathbf{R}}$ given by

$$h_+(\boldsymbol{d}) := \begin{cases} 0 & \text{if } \boldsymbol{d} \in \mathrm{cl}(\mathcal{D}_f^<(\boldsymbol{x}_0)) \\ f'_+(\boldsymbol{x}_0;\boldsymbol{d}) & \text{otherwise} \end{cases} .$$

If $\boldsymbol{d} \longmapsto f'_+(\boldsymbol{x}_0;\boldsymbol{d})$ is proper, evenly quasiconvex it follows independently of the fact whether $\mathcal{D}_f^<(\boldsymbol{x}_0)$ is empty ($\mathrm{cl}(\emptyset) = \emptyset$) or not that $\boldsymbol{d} \longmapsto h_+(\boldsymbol{d})$ is a proper, evenly quasiconvex and positively homogeneous function. Again by Theorem 3.5 of Ref. [78], or Lemma 2.7 of Ref. [35], we obtain for every $\boldsymbol{d} \in \mathbf{R}^s$ that

$$0 \le h_+(\boldsymbol{d}) = \sup\{k_+(\boldsymbol{x}^\star, \boldsymbol{x}^{\star\mathsf{T}}\boldsymbol{d}) : \boldsymbol{x}^\star \in \mathbf{R}^s\} \tag{5.8}$$

with

$$k_+(\boldsymbol{x}^\star, r) := \inf\{h_+(\boldsymbol{d}) : \boldsymbol{x}^{\star\mathsf{T}}\boldsymbol{d} \ge r,\ \boldsymbol{d} \in \mathbf{R}^s\} \ge 0. \tag{5.9}$$

The following result for k_+ is related to Lemma 5.2.4.

Lemma 5.2.8. *For each $\boldsymbol{x}^\star \in \mathbf{R}^s$ the function $r \longmapsto k_+(\boldsymbol{x}^\star, r)$ is nondecreasing, nonnegative and positively homogeneous.*

Proof: Obvious. □

The above result can be strengthen as follows.

Lemma 5.2.9.

1. *If $\boldsymbol{x}^\star$ does not belong to $(\mathrm{cl}(\mathcal{D}_f^<(\boldsymbol{x}_0)))^\circ$ then $k_+(\boldsymbol{x}^\star, r) = 0$ for every $r \in \mathbf{R}$.*
2. *If $\boldsymbol{x}^\star$ belongs to $(\mathrm{cl}(\mathcal{D}_f^<(\boldsymbol{x}_0)))^\circ$ then*

$$k_+(\boldsymbol{x}^\star, r) = \begin{cases} 0 & \textit{for every } r \le 0 \\ r k_+(\boldsymbol{x}^\star, 1) & \textit{for every } r > 0 \end{cases}$$

and $0 \le k_+(\boldsymbol{x}^\star, 1) \le \infty$.

Proof: Clearly $k_+(\boldsymbol{x}^\star, 0) = \inf\{h_+(\boldsymbol{d}) : \boldsymbol{x}^{\star\mathsf{T}}\boldsymbol{d} \geq 0,\ \boldsymbol{d} \in \mathbf{R}^s\} \leq h_+(\mathbf{0}) = 0$. By Lemma 5.2.8 we know that $r \longmapsto k_+(\boldsymbol{x}^\star, r)$ is nondecreasing and nonnegative and so $k_+(\boldsymbol{x}^\star, r) = 0$ for every $r \leq 0$. Suppose now that $\boldsymbol{x}^\star$ does not belong to $(\mathrm{cl}(\mathcal{D}_f^<(\boldsymbol{x}_0)))^\circ$ (remember that if $\mathcal{D}_f^<(\boldsymbol{x}_0)$ is empty then $(\mathrm{cl}(\mathcal{D}_f^<(\boldsymbol{x}_0)))^\circ = \mathbf{R}^s$). This means that $\mathcal{D}_f^<(\boldsymbol{x}_0)$ is nonempty and there exists some $\boldsymbol{d}_0 \in \mathcal{D}_f^<(\boldsymbol{x}_0)$ satisfying $\boldsymbol{x}^{\star\mathsf{T}}\boldsymbol{d}_0 > 0$. By Lemma 5.2.8 and (5.9) we obtain for every $r > 0$ that

$$\begin{aligned} k_+(\boldsymbol{x}^\star, r) &= (\boldsymbol{x}^{\star\mathsf{T}}\boldsymbol{d}_0)^{-1} k_+(\boldsymbol{x}^\star, r\boldsymbol{x}^{\star\mathsf{T}}\boldsymbol{d}_0) \\ &\leq (\boldsymbol{x}^{\star\mathsf{T}}\boldsymbol{d}_0)^{-1} h_+(r\boldsymbol{d}_0) = r(\boldsymbol{x}^{\star\mathsf{T}}\boldsymbol{d}_0)^{-1} h_+(\boldsymbol{d}_0) = 0 \end{aligned}$$

and so in this case $k_+(\boldsymbol{x}^\star, r) = 0$ for every $r \in \mathbf{R}$.

This proves 1. and to verify 2. we only have to observe by Lemma 5.2.8 that $k_+(\boldsymbol{x}^\star, r) = r k_+(\boldsymbol{x}^\star, 1)$ for every $r > 0$. □

Similar to the function h_- discussed in Lemma 5.2.7, one can show the following result for the function h_+. Again this result improves a related result in Ref. [23].

Lemma 5.2.10. *If $\boldsymbol{d} \longmapsto f'_+(\boldsymbol{x}_0; \boldsymbol{d})$ is proper and evenly quasiconvex and the function $h_+ : \mathbf{R}^s \longrightarrow \overline{\mathbf{R}}$ is given by*

$$h_+(\boldsymbol{d}) = \begin{cases} 0 & \text{if } \boldsymbol{d} \in \mathrm{cl}(\mathcal{D}_f^<(\boldsymbol{x}_0)) \\ f'_+(\boldsymbol{x}_0; \boldsymbol{d}) & \text{otherwise} \end{cases}$$

then $\boldsymbol{d} \longmapsto h_+(\boldsymbol{d})$ is a lower semicontinuous sublinear function and its subgradient set $\partial h_+(\mathbf{0})$ at $\mathbf{0}$ contains $\mathbf{0}$. Moreover, $\partial h_+(\mathbf{0})$ is compact if and only if the function $\boldsymbol{d} \longmapsto f'_+(\boldsymbol{x}_0; \boldsymbol{d})$ is finite for every $\boldsymbol{d} \in \mathbf{R}^s$.

Proof: By Lemma 5.2.9 it is easy to verify that $r \longmapsto k_+(\boldsymbol{x}^\star, r)$ is lower semicontinuous, convex and positively homogeneous for every $\boldsymbol{x}^\star \in \mathbf{R}^s$. Hence, also the function $\boldsymbol{d} \longmapsto k_+(\boldsymbol{x}^\star, \boldsymbol{x}^{\star\mathsf{T}}\boldsymbol{d})$ is lower semicontinuous, positively homogeneous and convex and so by (5.8) it follows that $\boldsymbol{d} \longmapsto h_+(\boldsymbol{d})$ is lower semicontinuous, positively homogeneous and convex. Since $\boldsymbol{d} \longmapsto f'_+(\boldsymbol{x}_0; \boldsymbol{d})$ is proper we obtain by the definition of h_+ that $\boldsymbol{d} \longmapsto h_+(\boldsymbol{d})$ is also proper and hence we have shown that $\boldsymbol{d} \longmapsto h_+(\boldsymbol{d})$ is lower semicontinuous and sublinear. By Theorem 2.3.5 it follows that $\partial h_+(\mathbf{0})$ is nonempty while by Corollary 2.3.2 this set is compact if and only if $d \longmapsto f'_+(x_0; d)$ is finite for every $d \in \mathbf{R}^s$. To verify that $\mathbf{0} \in \partial h_+(\mathbf{0})$ we observe since $\partial h_+(\mathbf{0})$ is closed that by Theorem III.4.1.1 of Ref. [52] there exists some $d_0 \in \mathbf{R}^s$ and $\varepsilon > 0$ such that $d_0^{\mathsf{T}} x \leq -\varepsilon$ for every $x \in \partial h_+(\mathbf{0})$ if $\mathbf{0}$ does not belong to $\partial h_+(\mathbf{0})$. Since h_+ is by Theorem 2.3.5 the support function of $\partial h_+(\mathbf{0})$ we obtain $h_+(d_0) \leq -\varepsilon$ and this contradicts $h_+(d) \geq 0$ for every $d \in \mathbf{R}^s$. □

An immediate consequence of Lemma 5.2.2 and Lemma 5.2.10 is given by the following result. Although this result is not difficult to prove it appears to be new.

Lemma 5.2.11. *If $d \longmapsto f'_+(x_0; d)$ is proper and evenly quasiconvex, $\mathcal{L}_f^{<}(f(x_0))$ nonempty and $\mathcal{D}_f^{<}(x_0)$ empty then* $\text{cl}(\text{cone}(\partial h_+(\mathbf{0}))) \subseteq \mathcal{N}_{\mathcal{L}_f^{<}}(x_0)$.

Proof: By our assumptions it follows that $\mathcal{D}_f^{=}(x_0)$ is nonempty and $\mathcal{D}_f^{\leq}(x_0) = \mathcal{D}_f^{=}(x_0)$. This implies by Lemma 5.2.10 and h_+ being the support function of $\partial h_+(\mathbf{0})$ that

$$\begin{aligned}\mathcal{D}_f^{=}(x_0) &= \{d \in \mathbf{R}^s : h_+(d) = 0\} \\ &= \{d \in \mathbf{R}^s : x^{\star\mathsf{T}} d \leq 0 \text{ for every } d \in \partial h_+(\mathbf{0})\} \\ &= (\text{cone}(\partial h_+(\mathbf{0})))^{\circ}.\end{aligned}$$

Applying now Lemma 5.2.2 and Lemma 2.2.2 yields

$$\text{cl}(\text{cone}(\partial h_+(\mathbf{0}))) = (\text{cone}(\partial h_+(\mathbf{0})))^{\circ\circ} = (\mathcal{D}_f^{=}(x_0))^{\circ} \subseteq \mathcal{N}_{\mathcal{L}_f^{<}}(x_0)$$

and so the desired result is proven. □

As already observed, if $d \longmapsto f'_+(x_0; d)$ is the zero functional, or equivalently if $\partial h_+(\mathbf{0}) = \{\mathbf{0}\}$, the above result does not provide any useful information. By Lemma 5.2.7 it follows for $d \longmapsto f'_+(x_0; d)$ proper and evenly quasiconvex and $\mathcal{D}_f^{<}(x_0)$ nonempty that $h_- : \mathbf{R}^s \longrightarrow \overline{\mathbf{R}}$ is lower semicontinuous and sublinear and this implies (as already observed in the proof of Lemma 5.2.7) that $\text{cl}(\mathcal{D}_f^{<}(x_0)) = \{d \in \mathbf{R}^s : h_-(d) \leq 0\}$. This yields, using $h_+(d) \geq 0$ for every $d \in \mathbf{R}^s$, that

$$f'_+(x_0; d) = \min\{h_-(d), h_+(d)\}. \tag{5.10}$$

Using the above relation we can derive the next result.

Lemma 5.2.12. *If* $d \longmapsto f'_+(x_0; d)$ *is a proper evenly quasiconvex function then* $d \longmapsto f'_+(x_0; d)$ *is lower semicontinuous.*

Proof: If $\mathcal{D}_f^<(x_0)$ is empty then by applying Lemma 5.2.10 we have $f'_+(x_0; d) = h_+(d)$ and h_+ is lower semicontinuous. Moreover, if $\mathcal{D}_f^<(x_0)$ is nonempty we obtain by (5.10) that $f'_+(x_0; d) = \min\{h_-(d), h_+(d)\}$ and the desired result follows by Lemma 5.2.7 and Lemma 5.2.10. □

By the above result, every proper evenly quasiconvex positively homogeneous function must be lower semicontinuous. This is a rather remarkable result which does not hold for evenly quasiconvex functions. As an example we mention the evenly quasiconvex function $x \longmapsto \text{sign}(x)$, defined in the beginning of Section 2.4, which is neither lower nor upper semicontinuous at 0.

If $d \longmapsto f'_+(x_0; d)$ is a finite evenly quasiconvex function one can show, under some additional conditions, that $d \longmapsto f'_+(x_0; d)$ is continuous on $\mathbb{R}^s$. Although these results are not important for the construction of a separation oracle we only list them since they appear to be new and they answer an obvious question by showing under which type of conditions there is no discontinuity gap on the boundary of $\mathcal{D}_f^<(x_0)$.

Lemma 5.2.13. *For* $\mathcal{D}_f^<(x_0)$ *nonempty and* $d \longmapsto f'_+(x_0; d)$ *proper evenly quasiconvex the following conditions are equivalent.*

1. $\text{rbd}(\mathcal{D}_f^<(x_0)) \subseteq \mathcal{D}_f^=(x_0)$.
2. $\mathcal{D}_f^<(x_0)$ *is relatively open.*

Proof: To verify 1. ⇒ 2. it is sufficient to prove that $\mathcal{D}_f^<(x_0) \subseteq \text{ri}(\mathcal{D}_f^<(x_0))$. Let $d_0 \in \mathcal{D}_f^<(x_0) \subseteq \text{cl}(\mathcal{D}_f^<(x_0))$ and suppose that d_0 does not belong to $\text{ri}(\mathcal{D}_f^<(x_0))$. Then $d_0 \in \text{rbd}(\mathcal{D}_f^<(x_0))$ and hence by 1. we obtain that $f'_+(x_0; d_0) = 0$. This contradicts $d_0 \in \mathcal{D}_f^<(x_0)$ and so we have shown that $\mathcal{D}_f^<(x_0) \subseteq \text{ri}(\mathcal{D}_f^<(x_0))$.

To prove 2. ⇒ 1. we observe for $d \in \text{rbd}(\mathcal{D}_f^<(x_0))$ that d does not belong to $\text{ri}(\mathcal{D}_f^<(x_0)) = \mathcal{D}_f^<(x_0)$. Hence, $f'_+(x_0; d) \geq 0$ and since $h_+(d) = 0$ for every

$\boldsymbol{d} \in \mathrm{cl}(\mathcal{D}_f^{<}(\boldsymbol{x}_0))$ it follows by (5.10) that $0 \leq f'_+(\boldsymbol{x}_0;\boldsymbol{d}) = \min\{h_-(\boldsymbol{d}), h_+(\boldsymbol{d})\} \leq 0$ or equivalently $f'_+(\boldsymbol{x}_0;\boldsymbol{d}) = 0$. □

If f is a quasiconvex function with $\mathcal{D}_f^{<}(\boldsymbol{x}_0)$ nonempty and f is continuous on a relative neighborhood of the point $\boldsymbol{x}_0 \in \mathrm{ri}(\mathrm{dom}(f))$ then one can show that 2. of Lemma 5.2.13 is satisfied. In fact, let $\boldsymbol{d} \in \mathcal{D}_f^{<}(\boldsymbol{x}_0)$. By our assumption we can find some $t_0 > 0$ such that $\boldsymbol{x}_0 + t_0\boldsymbol{d} \in \mathrm{ri}(\mathcal{L}_f^{<}(f(\boldsymbol{x}_0)))$ and so by Theorem 2.2.8 we obtain that $\boldsymbol{d} \in \frac{1}{t_0}(\mathrm{ri}(\mathcal{L}_f^{<}(f(\boldsymbol{x}_0))) - \boldsymbol{x}_0) \subseteq \mathrm{ri}(\mathrm{cone}(\mathcal{L}_f^{<}(f(\boldsymbol{x}_0)) - \boldsymbol{x}_0))$. Hence, by Lemma 5.2.3 it follows that $\boldsymbol{d}$ belongs to $\mathrm{ri}(\mathcal{D}_f^{<}(\boldsymbol{x}_0))$ and so $\mathcal{D}_f^{<}(\boldsymbol{x}_0)$ is relatively open. By the above lemma it is now possible to prove the following important result. Observe that an empty set is by definition relatively open.

Lemma 5.2.14. *If the function $\boldsymbol{d} \longmapsto f'_+(\boldsymbol{x}_0;\boldsymbol{d})$ is a finite evenly quasiconvex function and $\mathcal{D}_f^{<}(\boldsymbol{x}_0)$ is relatively open then the function $\boldsymbol{d} \longmapsto f'_+(\boldsymbol{x}_0;\boldsymbol{d})$ is continuous on $\mathbb{R}^s$.*

Proof: If $\mathcal{D}_f^{<}(\boldsymbol{x}_0)$ is empty then clearly by Lemma 5.2.10 we obtain that $f'_+(\boldsymbol{x}_0;\boldsymbol{d}) = h_+(\boldsymbol{d})$ for every $\boldsymbol{d} \in \mathbb{R}^s$. Since $\mathrm{dom}(h_+) = \mathbb{R}^s$ and again by Lemma 5.2.10 the function h_+ is convex it follows by Corollary 2.3.1 that $\boldsymbol{d} \longmapsto f'_+(\boldsymbol{x}_0;\boldsymbol{d})$ is continuous. If $\mathcal{D}_f^{<}(\boldsymbol{x}_0)$ is nonempty then by Theorem 2.3.1, Lemma 5.2.7, Lemma 5.2.10 and (5.10) it is sufficient to prove that $\boldsymbol{d} \longmapsto f'_+(\boldsymbol{x}_0;\boldsymbol{d})$ is upper semicontinuous on $\mathrm{rbd}(\mathcal{D}_f^{<}(\boldsymbol{x}_0))$. Since $\mathcal{D}_f^{<}(\boldsymbol{x}_0)$ is relatively open by assumption we obtain by Lemma 5.2.13 that $f'_+(\boldsymbol{x}_0;\boldsymbol{d}) = 0$ for every $\boldsymbol{d} \in \mathrm{rbd}(\mathcal{D}_f^{<}(\boldsymbol{x}_0))$. Suppose now by contradiction that $\limsup_{\boldsymbol{d}\to\boldsymbol{d}_0} f'_+(\boldsymbol{x}_0;\boldsymbol{d}) > f'_+(\boldsymbol{x}_0;\boldsymbol{d}_0) = 0$ for some $\boldsymbol{d}_0 \in \mathrm{rbd}(\mathcal{D}_f^{<}(\boldsymbol{x}_0))$. Hence, there exists a sequence $\{\boldsymbol{d}_k : k \geq 1\}$ with $\lim_{k\uparrow\infty} \boldsymbol{d}_k = \boldsymbol{d}_0$ such that $\lim_{k\uparrow\infty} f'_+(\boldsymbol{x}_0;\boldsymbol{d}_k) > 0$. By the continuity of h_+ and (5.10) this yields $h_+(\boldsymbol{d}_0) = \lim_{k\uparrow\infty} h_+(\boldsymbol{d}_k) \geq \lim_{k\uparrow\infty} f'_+(\boldsymbol{x}_0;\boldsymbol{d}_k) > 0$ and so $\boldsymbol{d}_0 \notin \mathrm{cl}(\mathcal{D}_f^{<}(\boldsymbol{x}_0))$ contradicting $\boldsymbol{d}_0 \in \mathrm{rbd}(\mathcal{D}_f^{<}(\boldsymbol{x}_0))$. Therefore $\boldsymbol{d} \longmapsto f'_+(\boldsymbol{x}_0;\boldsymbol{d})$ must be upper semicontinuous for every $\boldsymbol{d} \in \mathrm{rbd}(\mathcal{D}_f^{<}(\boldsymbol{x}_0))$ and this proves the desired result. □

It is now immediately clear for $\boldsymbol{d} \longmapsto f'_+(\boldsymbol{x}_0;\boldsymbol{d})$ continuous on $\mathbb{R}^s$ that the set $\mathcal{D}_f^{<}(\boldsymbol{x}_0)$, if not empty, has full dimension s and so $\mathrm{ri}(\mathcal{D}_f^{<}(\boldsymbol{x}_0)) = \mathrm{int}(\mathcal{D}_f^{<}(\boldsymbol{x}_0)) = \mathcal{D}_f^{<}(\boldsymbol{x}_0)$. Finally, if f is quasiconvex and locally Lipschitz around $\boldsymbol{x}_0$ (Ref. [19]) it is easy to show by a direct proof that $\boldsymbol{d} \longmapsto f'_+(\boldsymbol{x}_0;\boldsymbol{d})$ is Lipschitz continuous (and hence continuous) on $\mathbf{R}^s$. On the other hand, notice that Crouzeix (Ref. [21–23]) observed that $\boldsymbol{d} \longmapsto f'_+(\boldsymbol{x}_0; \boldsymbol{d})$ might not be lower semicontinuous even if $\boldsymbol{d} \longmapsto f'_+(\boldsymbol{x}_0; \boldsymbol{d})$ is a finite, positively homogeneous and quasiconvex function. This implies that the optimization problem $\inf_{\boldsymbol{d} \in \mathcal{C}} f'_+(\boldsymbol{x}_0; \boldsymbol{d})$ discussed in the next section with $\mathcal{C} \subseteq \mathbf{R}^s$ some compact convex set and $\mathbf{0} \in \text{int}(\mathcal{C})$ might not have a feasible optimal solution. However, as shown by the previous results, the set of proper, positively homogeneous and evenly quasiconvex functions coincides with the set of proper, positively homogeneous and lower semicontinuous quasiconvex functions and hence the concept of evenly quasiconvex is well suited for optimization purposes.

Using the previous results we discuss in the next subsection how to compute by means of an optimization problem an element of the normal cone $\mathcal{N}_{\mathcal{L}_f^<}(\boldsymbol{x}_0)$. This result, due to Frenk, Gromicho and Zhang in Ref. [39], seems to be new for quasiconvex functions and generalizes a similar result for convex functions.

5.2.2. Computing an element of the normal cone

In this subsection we analyze the problem of computing an element of the normal cone $\mathcal{N}_{\mathcal{L}_f^<}(\boldsymbol{x}_0)$ if $\mathcal{L}_f^<(f(\boldsymbol{x}_0))$ is nonempty. As already observed, a sufficient condition for $\mathcal{L}_f^<(f(\boldsymbol{x}_0))$ to be nonempty is given by the nonemptiness of the set $\mathcal{D}_f^<(\boldsymbol{x}_0)$ and so it is natural to consider the optimization problem

$$(S) \quad \inf\{f'_+(\boldsymbol{x}_0; \boldsymbol{d}) : \boldsymbol{d} \in \mathcal{C}\}$$

with $\mathcal{C}$ a compact convex set satisfying $\mathbf{0} \in \text{int}(\mathcal{C})$. Notice, since $\dim(\mathcal{C}) = s$ that the boundary $\text{bd}(\mathcal{C})$ of $\mathcal{C}$ is given by $\mathcal{C} \setminus \text{int}(\mathcal{C})$. In order to guarantee that (S) is solvable, i.e. there exists some $\boldsymbol{d}_0 \in \mathcal{C}$ with $f'_+(\boldsymbol{x}_0; \boldsymbol{d}_0) = \vartheta(S)$ and $\vartheta(S) := \inf\{f'_+(\boldsymbol{x}_0; \boldsymbol{d}) : \boldsymbol{d} \in \mathcal{C}\} \leq 0$, it is sufficient by Lemma 5.2.12 to assume that $\boldsymbol{d} \longmapsto f'_+(\boldsymbol{x}_0; \boldsymbol{d})$ is a proper evenly quasiconvex function. In the remainder of this section we always assume that $\boldsymbol{d} \longmapsto f'_+(\boldsymbol{x}_0; \boldsymbol{d})$ satisfies this property and so the set $\mathcal{S}$ of optimal solutions of (S) is always nonempty. Clearly, one should choose the compact convex set $\mathcal{C}$ with $\mathbf{0} \in \text{int}(\mathcal{C})$ in such a way that (S) is "easy" solvable. Since $\boldsymbol{d} \longmapsto f'_+(\boldsymbol{x}_0; \boldsymbol{d})$ is a proper, positively homogeneous and evenly quasiconvex function the following result is easy to verify and so its proof is omitted.

Lemma 5.2.15. *$\vartheta(S) < 0$ if and only if $\mathcal{D}_f^{<}(\boldsymbol{x}_0)$ is nonempty. If this holds then $\mathcal{S} \subseteq \mathrm{bd}(\mathcal{C})$. Moreover, if $\vartheta(S) = 0$, i.e. $\boldsymbol{x}_0 \in \Gamma_f$, then either $\mathbf{0}$ is the unique solution of (S) or $\mathcal{S} \cap \mathrm{bd}(\mathcal{C})$ is nonempty.*

Clearly, if $\mathbf{0}$ is the unique solution of optimization problem (S) then due to $\mathbf{0} \in \mathrm{int}(\mathcal{C})$, $f : \mathbf{R}^s \longrightarrow \overline{\mathbf{R}}$ quasiconvex and $\boldsymbol{d} \longmapsto f'_+(\boldsymbol{x}_0; \boldsymbol{d})$ proper and positively homogeneous, it follows that $\mathcal{L}_f^{<}(f(\boldsymbol{x}_0))$ is empty. Also, if $\vartheta(S) < 0$ we obtain by Lemma 5.2.7 that an optimal solution $\boldsymbol{d}_0$ of (S) is also an optimal solution of the optimization problem

$$(S') \quad \min\{h_-(\boldsymbol{d}) : \gamma_{\mathcal{C}}(\boldsymbol{d}) \leq 1,\ \boldsymbol{d} \in \mathrm{cl}(\mathcal{D}_f^{<}(\boldsymbol{x}_0))\}$$

with $\gamma_{\mathcal{C}}(\boldsymbol{d}) := \inf\{t > 0 : \boldsymbol{d} \in t\mathcal{C}\}$ the *gauge* of $\mathcal{C}$. Since by Lemma 5.2.7 the function h_- is proper and convex with $\mathrm{dom}(h_-) = \mathrm{cl}(\mathcal{D}_f^{<}(\boldsymbol{x}_0))$ and the function $\gamma_{\mathcal{C}}$ is finite and convex due to $\mathcal{C}$ compact, convex and $\mathbf{0} \in \mathrm{int}(\mathcal{C})$, the optimization problem (S') satisfies the properties of a convex program given in Section 28 of Ref. [89].

Using now so-called *primal-dual* information given by the *Karush-Kuhn-Tucker* conditions it is possible to prove the next result.

Lemma 5.2.16. *If $\boldsymbol{d}_0$ is an optimal solution of (S) with $\vartheta(S) < 0$ then the set $\partial h_-(\boldsymbol{d}_0) - \vartheta(S)\partial\gamma_{\mathcal{C}}(\boldsymbol{d}_0)$ contains $\mathbf{0}$.*

Proof: If $\boldsymbol{d}_0$ is an optimal solution of (S) with $f'_+(\boldsymbol{x}_0; \boldsymbol{d}_0) = \vartheta(S) < 0$ then $\lambda\boldsymbol{d}_0 \in \mathcal{D}_f^{<}(\boldsymbol{x}_0)$ for every $\lambda > 0$. Also, by Lemma 5.2.15 we obtain that $\boldsymbol{d}_0 \in \mathrm{bd}(\mathcal{C})$ implying that $\gamma_{\mathcal{C}}(\boldsymbol{d}_0) = 1$ and so $\gamma_{\mathcal{C}}(\lambda\boldsymbol{d}_0) = \lambda\gamma_{\mathcal{C}}(\boldsymbol{d}_0) < 1$ for every $0 < \lambda < 1$. Hence, by Corollary 28.2.1 of Ref. [89] a Karush-Kuhn-Tucker vector λ_1 of (S') exists and this yields by Theorem 28.3 of Ref. [89] that $\mathbf{0} \in \partial h_-(\boldsymbol{d}_0) + \lambda_1\partial\gamma_{\mathcal{C}}(\boldsymbol{d}_0)$, $\lambda_1(\gamma_{\mathcal{C}}(\boldsymbol{d}_0) - 1) = 0$ and $\lambda_1 \geq 0$. If $\lambda_1 = 0$ it follows that $\mathbf{0} \in \partial h_-(\boldsymbol{d}_0)$ and this yields $h_-(\boldsymbol{d}) \geq h_-(\boldsymbol{d}_0)$ for every $\boldsymbol{d} \in \mathbf{R}^s$. However, since h_- is positively homogeneous and $h_-(\boldsymbol{d}_0) < 0$ it follows that $h_-(\lambda\boldsymbol{d}_0) = \lambda h_-(\boldsymbol{d}_0) < h_-(\boldsymbol{d}_0)$ for every $\lambda > 1$ contradicting $\mathbf{0} \in \partial h_-(\boldsymbol{d}_0)$. Hence, $\lambda_1 > 0$ and to compute λ_1 we observe the following. It is well-known (Theorem 2.3.8) that $\partial\gamma_{\mathcal{C}}(\boldsymbol{d}_0) = \{\boldsymbol{d}_0^\star \in \mathcal{C}^\circ : \boldsymbol{d}_0^{\star\intercal}\boldsymbol{d}_0 = \gamma_{\mathcal{C}}(\boldsymbol{d}_0)\}$ with $\mathcal{C}^\circ$ the polar of $\mathcal{C}$ and so by the Karush-Kuhn-Tucker conditions and Lemma 5.2.15 there exists some $\boldsymbol{d}_0^\star \in \mathbf{R}^s$ with $-\boldsymbol{d}_0^\star \in \partial h_-(\boldsymbol{d}_0)$, $\boldsymbol{d}_0^{\star\intercal}\boldsymbol{d}_0 = \lambda_1$ and

$d_0^{\star T} d \leq \lambda_1$ for every $d \in \mathcal{C}$. Since $-d_0^\star \in \partial h_-(d_0)$ it follows by Theorem 23.5 of Ref. [89] that $h_-(d_0) + h_-^*(-d_0^\star) = -d_0^{\star T} d_0$ with h_-^* the conjugate function of h_-. Since h_- is positively homogeneous and thus h_-^* is either 0 or $+\infty$ we obtain by the above equality that $h_-(d_0) = -d_0^{\star T} d_0 = -\lambda_1$ and so the result is proven. □

The following result is an immediate consequence of the previous lemma.

Corollary 5.2.1. *If γ_C is differentiable in d_0 then $-\nabla\gamma_C(d_0) \in \mathcal{N}_{\mathcal{L}_f^<}(x_0)$.*

Proof: The previous result shows for $\vartheta(S) < 0$ and d_0 an optimal solution of (S) that the sets $\partial h_-(d_0)$ and $\vartheta(S)\partial\gamma_C(d_0)$ intersect. Hence, if γ_C is differentiable in d_0 with gradient $\nabla\gamma_C(d_0)$ then $\vartheta(S)\partial\gamma_C(d_0) = \{\vartheta(S)\nabla\gamma_C(d_0)\}$ and so $\vartheta(S)\nabla\gamma_C(d_0) \in \partial h_-(d_0) \subseteq \partial h_-(0)$. Now, by Lemma 5.2.7 it follows that $\partial h_-(0) \subseteq \mathcal{N}_{\mathcal{L}_f^<}(x_0)$ and since $\vartheta(S) < 0$ and $\mathcal{N}_{\mathcal{L}_f^<}(x_0)$ is a cone this leads to the stated result. □

Example 5.2.1. *Take $\mathcal{C} := \{x \in \mathbb{R}^s : \|x\|_p \leq 1\}$ with $\|\cdot\|_p$, $1 < p < \infty$, the ℓ_p-norm. Clearly, $\gamma_C(x) = \|x\|_p$ and γ_C is differentiable everywhere except at $\mathbf{0}$. Moreover, for every $x \neq \mathbf{0}$ it is easy to verify that*

$$\nabla\gamma_C(x) = \|x\|_p^{1-p} \begin{bmatrix} \operatorname{sign}(x_1)|x_1|^{p-1} \\ \vdots \\ \operatorname{sign}(x_s)|x_s|^{p-1} \end{bmatrix}$$

with x_i the ith component of x and $\operatorname{sign}(x)$ the sign function.

If $\vartheta(S) < 0$, d_0 solves (S) and γ_C is not differentiable in d_0 while $d \longmapsto f'_+(x_0; d)$ is differentiable in d_0 then it is easy to show, due to $d_0 \in \operatorname{int}(\mathcal{D}_f^<(x_0))$ and the definition of h_-, that $\nabla h(d_0) = \nabla h_-(d_0) \in \partial h_-(0)$ with $h(d) := f'_+(x_0; d)$. In this case, the optimization problem (S) is only used to identify an interior element of $\mathcal{D}_f^<(x_0)$. This also shows that selecting some $d_1 \in \operatorname{int}(\mathcal{D}_f^<(x_0))$ with h differentiable in d_1 already yields an element of the normal cone $\mathcal{N}_{\mathcal{L}_f^<}(x_0)$.

Finally, we provide in Figure 5.5 a geometrical interpretation of Lemma 5.2.16. The first picture shows a set $\mathcal{C}$ with a kink at d_0 and for which $\operatorname{cone}(\partial\gamma_C(d_0))$ is a cone (shifted in the picture to the vertex of C for clarity) whose symmetric cone intersects $\partial h_{-}(\boldsymbol{d}_0)$ (by Lemma 5.2.16) but includes elements which do not belong to $\text{cone}(\partial h_{-}(\boldsymbol{d}_0)) = \mathcal{N}_{\mathcal{L}_f^<}(\boldsymbol{x}_0)$.

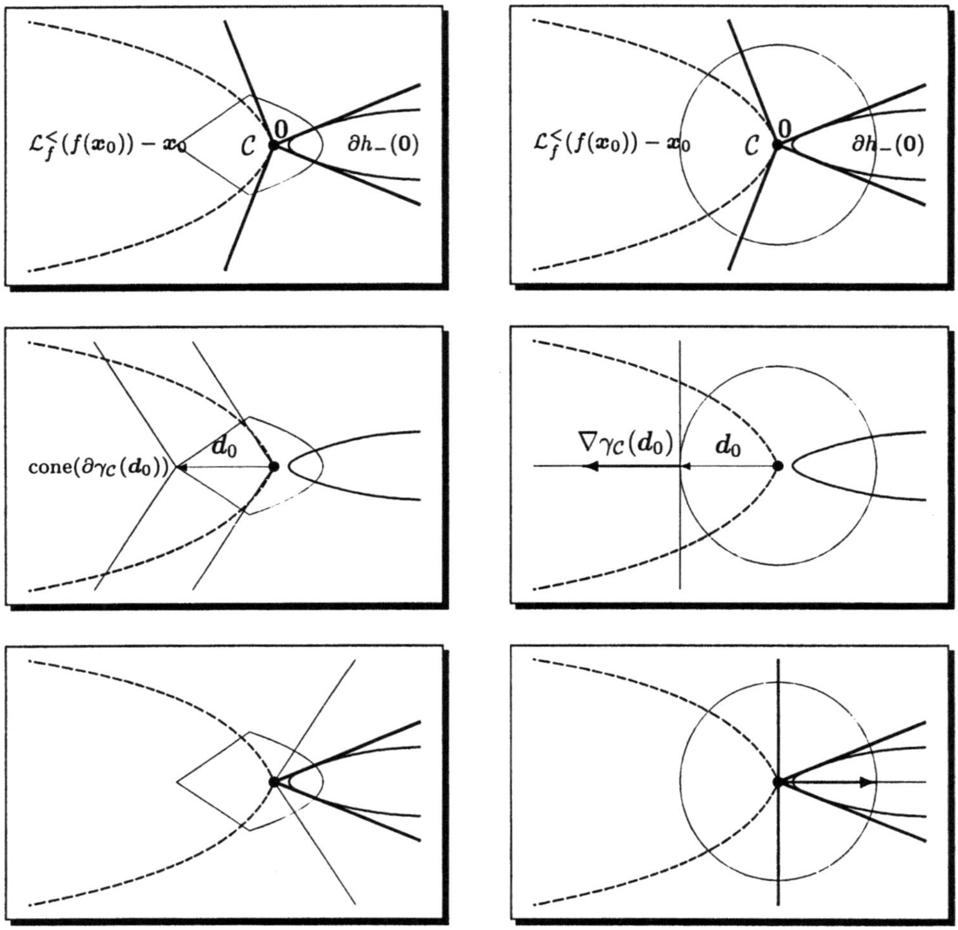

Figure 5.5: Geometric interpretation of the separation oracle

On the other hand, the second picture corresponds to a smooth C. Hence $\partial\gamma_C(\boldsymbol{d}_0)$ is a singleton and so the symmetric of its conical hull (a half line) intersects (by Lemma 5.2.16 again) $\text{cone}(\partial h_{-}(\boldsymbol{d}_0))$ and it must be included in $\mathcal{N}_{\mathcal{L}_f^<}(\boldsymbol{x}_0)$.

We triplicate each picture for clarity. The top picture shows the conical hull of the strict lower level set, the corresponding normal cone and the compact convex set C

corresponding to the feasible region of problem (S). The middle picture shows the solution of problem (S), direction $\boldsymbol{d}_0$, and the conical hull of $\partial\gamma_C(\boldsymbol{d}_0)$. Finally, the bottom picture shows the intersection of the symmetric of this conical hull and the normal cone. We consider in the next section a class of quasiconvex functions and another example for which we do not have to solve the optimization problem (S).

5.3. Easy cases

This section illustrates classes of functions for which the optimization problem (S) can be replaced by easier membership problems.

5.3.1. Regular functions

In this subsection we discuss a separation oracle for the following subclass of quasiconvex functions.

Definition 5.3.1. *Let $f : \mathbb{R}^s \longrightarrow \overline{\mathbb{R}}$ be a quasiconvex function with $f(\boldsymbol{x}_0)$ finite. Then f is called regular at $\boldsymbol{x}_0$ if $\boldsymbol{d} \longmapsto f'_+(\boldsymbol{x}_0;\boldsymbol{d})$ is a lower semicontinuous sublinear function.*

Following Pshenichnyi in Ref. [87] these functions are sometimes called *quasidifferentiable.* However, we prefer to follow Clarke (Ref. [19]) and call them *regular* since the term quasidifferentiable has nowadays a broader meaning (Ref. [25]).

As the next lemma shows the above class of functions is closed under the finite max operator.

Lemma 5.3.1. *Let $f_i : \mathbb{R}^s \longrightarrow \overline{\mathbb{R}}$, $i = 1, \ldots, n$, be quasiconvex functions with $f_i(\boldsymbol{x}_0)$ finite for every $1 \leq i \leq n$. If each function f_i is regular at $\boldsymbol{x}_0$ then the function $f : \mathbb{R}^s \longrightarrow \overline{\mathbb{R}}$ given by $f(\boldsymbol{x}) := \max_{1 \leq i \leq n} f_i(\boldsymbol{x})$ is also regular at $\boldsymbol{x}_0$.*

Proof: It is easy to verify that f is quasiconvex and $f(\boldsymbol{x}_0)$ is finite. Moreover, by Lemma 2.5.3 we obtain that

$$f'_+(\boldsymbol{x}_0;\boldsymbol{d}) = \max_{i \in I(\boldsymbol{x}_0)} f'_{i+}(\boldsymbol{x}_0;\boldsymbol{d}) \tag{5.11}$$

with $I(\boldsymbol{x}_0) := \{1 \leq i \leq n : f(\boldsymbol{x}_0) = f_i(\boldsymbol{x}_0)\}$ the set of active indices of f at $\boldsymbol{x}_0$. Since by assumption it follows that $\boldsymbol{d} \longmapsto f'_{i+}(\boldsymbol{x}_0; \boldsymbol{d})$ is a lower semicontinuous sublinear function for every $1 \leq i \leq n$ the desired result follows by (5.11). □

An important class of regular functions is given by the next lemma.

Lemma 5.3.2. *Let $g : \mathbb{R}^m \longrightarrow \mathbb{R}$ be a finite nondecreasing quasiconvex function and $\boldsymbol{v} : \mathbb{R}^s \longrightarrow \mathbb{R}^m$ a finite-valued convex vector function, i.e. $\boldsymbol{v}(\boldsymbol{x}) := (v_1(\boldsymbol{x}), \ldots, v_m(\boldsymbol{x}))$ with $v_i : \mathbb{R}^s \longrightarrow \mathbb{R}$, $i = 1, \ldots, m$, finite-valued convex functions. If the function $f : \mathbb{R}^s \longrightarrow \overline{\mathbb{R}}$ is given by $f(\boldsymbol{x}) = g(\boldsymbol{v}(\boldsymbol{x}))$ and g is regular and locally Lipschitz at $\boldsymbol{v}(\boldsymbol{x}_0)$ then f is a quasiconvex function regular at $\boldsymbol{x}_0$. Moreover, $f'_+(\boldsymbol{x}_0; \boldsymbol{d}) = g'_+(\boldsymbol{v}(\boldsymbol{x}_0); \boldsymbol{v}'(\boldsymbol{x}_0; \boldsymbol{d}))$ with $\boldsymbol{v}'(\boldsymbol{x}_0; \boldsymbol{d}) = (v'_1(\boldsymbol{x}_0; \boldsymbol{d}), \ldots, v'_m(\boldsymbol{x}_0; \boldsymbol{d}))$ and the function $\boldsymbol{d} \longmapsto f'_+(\boldsymbol{x}_0; \boldsymbol{d})$ is Lipschitz continuous.*

Proof: Since g is a nondecreasing quasiconvex function and $\boldsymbol{v}$ a convex vector function it is easy to verify that f is quasiconvex. Also, by Lemma 2.5.2 it follows that $f'_+(\boldsymbol{x}_0; \boldsymbol{d}) = g'_+(\boldsymbol{v}(\boldsymbol{x}_0); \boldsymbol{v}'(\boldsymbol{x}_0; \boldsymbol{d}))$. Moreover, since g is regular at $\boldsymbol{v}(\boldsymbol{x}_0)$ and nondecreasing we obtain for every $0 < \lambda < 1$ and $\boldsymbol{d}_1, \boldsymbol{d}_2 \in \mathbb{R}^s$ that

$$\begin{aligned}
& f'_+(\boldsymbol{x}_0; \lambda \boldsymbol{d}_1 + (1-\lambda)\boldsymbol{d}_2) \\
&= g'_+(\boldsymbol{v}(\boldsymbol{x}_0); \boldsymbol{v}'(\boldsymbol{x}_0; \lambda \boldsymbol{d}_1 + (1-\lambda)\boldsymbol{d}_2)) \\
&\leq g'_+(\boldsymbol{v}(\boldsymbol{x}_0); \lambda \boldsymbol{v}'(\boldsymbol{x}_0; \boldsymbol{d}_1) + (1-\lambda)\boldsymbol{v}'(\boldsymbol{x}_0; \boldsymbol{d}_2)) \\
&\leq \lambda g'_+(\boldsymbol{v}(\boldsymbol{x}_0); \boldsymbol{v}'(\boldsymbol{x}_0; \boldsymbol{d}_1)) + (1-\lambda) g'_+(\boldsymbol{v}(\boldsymbol{x}_0); \boldsymbol{v}'(\boldsymbol{x}_0; \boldsymbol{d}_2)) \\
&= \lambda f'_+(\boldsymbol{x}_0; \boldsymbol{d}_1) + (1-\lambda) f'_+(\boldsymbol{x}_0; \boldsymbol{d}_2)
\end{aligned}$$

and by the Lipschitz continuity of $\boldsymbol{d} \longmapsto \boldsymbol{v}'(\boldsymbol{x}_0; \boldsymbol{d})$ and $\boldsymbol{d} \longmapsto g'_+(\boldsymbol{v}(\boldsymbol{x}_0); \boldsymbol{d})$, the Lipschitz continuity of $\boldsymbol{d} \longmapsto f'_+(\boldsymbol{x}_0; \boldsymbol{d})$ follows. □

For the class of functions given in Definition 5.3.1 it is now easy, using only classical results of convex analysis, to prove the next result.

Lemma 5.3.3. *If $f : \mathbb{R}^s \longrightarrow \overline{\mathbb{R}}$ is a quasiconvex function regular at $\boldsymbol{x}_0$ and $\mathcal{D}_f^<(\boldsymbol{x}_0)$ is nonempty then* $\text{cl}(\text{cone}(\partial h(\mathbf{0}))) = \mathcal{N}_{\mathcal{L}_f^<}(\boldsymbol{x}_0)$ *with $\partial h(\mathbf{0})$ the subgradient set of the convex function $h(\boldsymbol{d}) := f'_+(\boldsymbol{x}_0; \boldsymbol{d})$ at $\mathbf{0}$. Moreover, if $\mathcal{L}_f^<(f(\boldsymbol{x}_0))$ is nonempty and $\mathcal{D}_f^<(\boldsymbol{x}_0)$ is empty then* $\text{cl}(\text{cone}(\partial h(\mathbf{0}))) \subseteq \mathcal{N}_{\mathcal{L}_f^<}(\boldsymbol{x}_0)$.

Proof: To prove the first result we observe by (5.3) that $\mathcal{N}_{\mathcal{L}_f^<}(x_0) = (\mathrm{cl}(\mathcal{D}_f^<(x_0)))^\circ$. Since f is a quasiconvex function regular at x_0 it follows by Lemma 2.3.1 that $\mathrm{cl}(\mathcal{D}_f^<(x_0)) = \mathcal{D}_f^\leq(x_0)$. Moreover, by Theorem 2.3.5 we obtain that $\mathcal{D}_f^\leq(x_0)$ equals $\{d : x^{\star\top} d \leq 0 \text{ for every } x^\star \in \partial h(0)\}$. Clearly this set also equals $(\mathrm{cone}(\partial h(0)))^\circ$ and hence by (5.3) and Lemma 2.2.2 we obtain that

$$\mathcal{N}_{\mathcal{L}_f^<}(x_0) = (\mathrm{cone}(\partial h(0)))^{\circ\circ} = \mathrm{cl}(\mathrm{cone}(\partial h(0))).$$

The second result can be proved in a similar way and this completes the proof. □

Finally we can show the main result of this subsection.

Lemma 5.3.4. *Let $g_i : \mathbb{R}^m \longrightarrow \mathbb{R}$, $i = 1, \ldots, n$, be quasiconvex and continuously differentiable functions and suppose $v_i : \mathbb{R}^s \longrightarrow \mathbb{R}$, $1 \leq i \leq m$, are finite-valued convex functions. Then it follows for $f(x) := \max_{1 \leq i \leq n} f_i(x)$ with $f_i(x) := g_i(v(x))$ that x_0 belongs to Γ_f if and only if*

$$0 \in \mathrm{conv}\left(\bigcup_{i \in I(x_0)} \sum_{j=1}^{m} \tfrac{\partial g_i}{\partial z_j}(v(x_0)) \partial v_j(x_0) \right)$$

where $I(x_0) := \{1 \leq i \leq n : f(x_0) = f_i(x_0)\}$. Moreover, if $\mathcal{D}_f^<(x_0)$ is nonempty, i.e. $x_0 \notin \Gamma_f$, then

$$\mathrm{cone}\left(\mathrm{conv}\left(\bigcup_{i \in I(x_0)} \sum_{j=1}^{m} \tfrac{\partial g_i}{\partial z_j}(v(x_0)) \partial v_j(x_0) \right) \right) = \mathcal{N}_{\mathcal{L}_f^<}(x_0)$$

Proof: Clearly every function $g_i : \mathbb{R}^m \longrightarrow \mathbb{R}$ is regular and locally Lipschitz at x_0. Applying now Lemma 5.3.2 yields

$$f_i'(x_0; d) = \sum_{j=1}^{m} \tfrac{\partial g_i}{\partial z_j}(v(x_0)) v_j'(x_0; d)$$

and so by Theorem 2.3.5 we obtain for $1 \leq i \leq n$ that

$$f_i'(x_0; d) = \max\{d^\top x^\star : x^\star \in \partial h_i(0)\}$$

with $\partial h_i(\mathbf{0}) := \sum_{j=1}^{m} \frac{\partial g_i}{\partial z_j}(\boldsymbol{v}(\boldsymbol{x}_0))\partial v_j(\boldsymbol{x}_0)$. By Lemma 2.5.3 this implies

$$\begin{aligned} f'(\boldsymbol{x}_0;\boldsymbol{d}) &= \max_{i\in I(\boldsymbol{x}_0)} f_i'(\boldsymbol{x}_0;\boldsymbol{d}) \\ &= \max_{i\in I(\boldsymbol{x}_0)} \max\{\boldsymbol{d}^{\intercal}\boldsymbol{x}^\star : \boldsymbol{x}^\star \in \partial h_i(\mathbf{0})\} \\ &= \max\left\{\boldsymbol{d}^{\intercal}\boldsymbol{x}^\star : \boldsymbol{x}^\star \in \operatorname{conv}\left(\bigcup_{i\in I(\boldsymbol{x}_0)} \partial h_i(\mathbf{0})\right)\right\}. \end{aligned}$$

Using the above relation it follows that $f'(\boldsymbol{x}_0;\boldsymbol{d}) \geq 0$ for every $\boldsymbol{d} \in \mathbb{R}^s$ if and only if $\mathbf{0}$ belongs to $\operatorname{conv}\left(\bigcup_{i\in I(\boldsymbol{x}_0)} \partial h_i(\mathbf{0})\right)$. This proves the first part. To prove the second part we observe that $\operatorname{conv}\left(\bigcup_{i\in I(\boldsymbol{x}_0)} \partial h_i(\mathbf{0})\right)$ is the subgradient set of the finite valued convex function $\boldsymbol{d} \longmapsto f'(\boldsymbol{x}_0;\boldsymbol{d})$ at $\mathbf{0}$ and since $\partial h_i(\mathbf{0})$ is compact for each i and $\mathbf{0}$ does not belong to $\operatorname{conv}\left(\bigcup_{i\in I(\boldsymbol{x}_0)} \partial h_i(\mathbf{0})\right)$ the second result follows by Lemma 5.3.3 together with Lemma 2.2.1 and Theorem 2.2.2. □

In the next subsection we consider another class of quasiconvex functions for which an "easy" separation oracle is available.

5.3.2. Another class of easy functions

Let $g_i : \mathbb{R}^s \longrightarrow \mathbb{R}$ be a continuously differentiable and convex function and $\gamma_i \in \mathbb{R}$, $1 \leq i \leq m$, and introduce the functions $f_i : \mathbb{R}^s \longrightarrow \mathbb{R}$ given by $f_i(\boldsymbol{x}) := \min\{g_i(\boldsymbol{x}), \gamma_i\}$. Clearly, the functions f_i are quasiconvex and so is the function $f(\boldsymbol{x}) := \max_{1\leq i\leq m} f_i(\boldsymbol{x})$. A representation of such a function for the case of affine g_i is given in Figure 5.6.

Moreover, if $I(\boldsymbol{x}_0) := \{1 \leq i \leq m : f(\boldsymbol{x}_0) = f_i(\boldsymbol{x}_0)\}$ it follows that

$$f'(\boldsymbol{x}_0;\boldsymbol{d}) = \max_{i\in I(\boldsymbol{x}_0)} f_i'(\boldsymbol{x}_0;\boldsymbol{d}) \tag{5.12}$$

and

$$f_i'(\boldsymbol{x}_0;\boldsymbol{d}) = \begin{cases} \min\{\nabla g_i(\boldsymbol{x}_0)^{\intercal}\boldsymbol{d}, 0\} & \text{if } g_i(\boldsymbol{x}_0) = \gamma_i \\ \nabla g_i(\boldsymbol{x}_0)^{\intercal}\boldsymbol{d} & \text{if } g_i(\boldsymbol{x}_0) < \gamma_i \\ 0 & \text{if } g_i(\boldsymbol{x}_0) > \gamma_i \end{cases}. \tag{5.13}$$

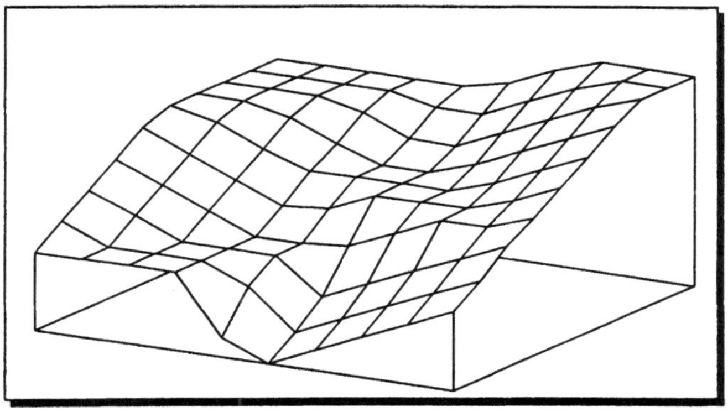

Figure 5.6: A bivariate quasiconvex function with horizontal regions

By (5.12) and (5.13) it is clear that $f'(\boldsymbol{x}_0; \boldsymbol{d}) \geq 0$ for every $\boldsymbol{d} \in \mathbb{R}^s$ if there exists some $i \in I(\boldsymbol{x}_0)$ satisfying $g_i(\boldsymbol{x}_0) > \gamma_i$ and this implies that $\vartheta(S) = \min_{\boldsymbol{d} \in C} f'(\boldsymbol{x}_0; \boldsymbol{d}) \geq 0$. Therefore assume for every $i \in I(\boldsymbol{x}_0)$ that $g_i(\boldsymbol{x}_0) \leq \gamma_i$. If this holds the following result is easy to prove.

Lemma 5.3.5. *If for every $i \in I(\boldsymbol{x}_0)$ it follows that $g_i(\boldsymbol{x}_0) \leq \gamma_i$ then $\vartheta(S) < 0$ if and only if $\boldsymbol{0} \notin \text{conv}(\{\nabla g_i(\boldsymbol{x}_0), i \in I(\boldsymbol{x}_0)\})$. Moreover, if this holds then $\text{conv}(\{\nabla g_i(\boldsymbol{x}_0), i \in I(\boldsymbol{x}_0)\}) \subseteq \mathcal{N}_{\mathcal{L}_f^{<}}(\boldsymbol{x}_0)$.*

Proof: Clearly, by the assumption $g_i(\boldsymbol{x}_0) \leq \gamma_i$ for every $i \in I(\boldsymbol{x}_0)$, (5.12) and (5.13) we obtain that $\min_{\boldsymbol{d} \in C} f'(\boldsymbol{x}; \boldsymbol{d})$ is equivalent to the optimization problem

$$\begin{array}{llll} \min & t & & \\ \text{st}: & t & \geq & \min\{\nabla g_i(\boldsymbol{x}_0)^{\top}\boldsymbol{d}, 0\} \quad \text{for every } i \in J(\boldsymbol{x}_0) \\ & t & \geq & \nabla g_i(\boldsymbol{x}_0)^{\top}\boldsymbol{d} \qquad\qquad\quad \text{for every } i \in I(\boldsymbol{x}_0) \setminus J(\boldsymbol{x}_0) \\ & \boldsymbol{d} & \in & C \end{array}$$

with $J(\boldsymbol{x}_0) := \{i \in I(\boldsymbol{x}_0) : g_i(\boldsymbol{x}_0) = \gamma_i\}$. This implies that $\vartheta(S) < 0$ if and only if the optimization problem

$$\begin{array}{llll} \min & t & & \\ \text{st}: & t & \geq & \nabla g_i(\boldsymbol{x}_0)^{\top}\boldsymbol{d} \quad \text{for every } i \in I(\boldsymbol{x}_0) \\ & \boldsymbol{d} & \in & C \end{array}$$

has a negative objective value. This problem in turn is equivalent to

$$\min_{d \in \mathcal{C}} \max_{i \in I(\boldsymbol{x}_0)} \nabla g_i(\boldsymbol{x}_0)^\intercal \boldsymbol{d} = \min_{d \in \mathcal{C}} \varphi(\boldsymbol{d})$$

with

$$\varphi(\boldsymbol{d}) := \max\{\boldsymbol{d}^\intercal \boldsymbol{y} : \boldsymbol{y} \in \operatorname{conv}(\{\nabla g_i(\boldsymbol{x}_0), i \in I(\boldsymbol{x}_0)\})\}.$$

We finally obtain that $\vartheta(S) < 0$ if and only if there exists some $\boldsymbol{d} \in \mathcal{C}$ with $\varphi(\boldsymbol{d}) < 0$ or equivalently $\boldsymbol{0} \notin \operatorname{conv}(\{\nabla g_i(\boldsymbol{x}_0), i \in I(\boldsymbol{x}_0)\})$. Observe that for $s = 2$ this decision can be carried out by means of Algorithm 4.2. By the definition of h_- and the representation of $f'(\boldsymbol{x}_0; \boldsymbol{d})$ it follows that

$$h_-(\boldsymbol{d}) = \begin{cases} \varphi(\boldsymbol{d}) & \text{if } \boldsymbol{d} \in \operatorname{cl}(\mathcal{D}_f^<(\boldsymbol{x}_0)) \\ +\infty & \text{otherwise} \end{cases}$$

and so any $\nabla g_i(\boldsymbol{x}_0)$, $i \in I(\boldsymbol{x}_0)$, belongs to $\partial h_-(\boldsymbol{0})$. This implies $\operatorname{conv}(\{\nabla g_i(\boldsymbol{x}_0), i \in I(\boldsymbol{x}_0)\}) \subseteq \partial h_-(\boldsymbol{0})$ and by Lemma 5.2.7 the desired result follows. □

Observe that if $\vartheta(S) = 0$ then we do not have enough information to proceed. Hence the next section discusses strategies to apply when no strict descent directions exist at $\boldsymbol{x}_0$.

5.4. When we meet a "bad" point

The previous sections show that there is no guarantee that the oracle is able to compute a nonzero element of the normal cone $\mathcal{N}_{\mathcal{L}_f^<}(\boldsymbol{x}_0)$ for $\boldsymbol{x}_0$ nonoptimal. In fact, if $\vartheta(S) = 0$ then the local information given by the Dini upper derivative is not sufficient to proceed. As Figure 5.6 shows, it is likely that such a "bad" point is encountered during the execution of the ellipsoid algorithm.

As introduced in (5.1) we denote the set of such points by Γ_f. Therefore suppose that at iteration m of the ellipsoid algorithm $\boldsymbol{a}_m$ belongs to Γ_f and so no separating hyperplane can be computed using the technique described in the previous section. To deal with this problem, the following results are extremely useful.

Lemma 5.4.1. *Suppose that $a_m \in \Gamma_f$. If a point $y \in \mathcal{L}_f^<(f(a_m)) \cap \mathcal{L}_{\bar{g}}^{\leq}(0) \cap \mathcal{L}_{\bar{n}}^{\leq}(r)$ exists then there also exists some $z_m \in \mathcal{L}_f^<(f(a_m)) \cap \mathcal{L}_{\bar{g}}^{\leq}(0) \cap \mathcal{L}_{\bar{n}}^{\leq}(r)$ satisfying $f'_+(z_m; z_m - a_m) < 0$. Moreover, this point z_m belongs to $]y, a_m[$.*

Proof: Since f is continuous and $f(y) < f(a_m)$ there exists some $0 < \eta < 1$ such that $w := \eta y + (1-\eta)a_m \in]y, a_m[$ satisfies $f(y) < f(w) = (f(y) + f(a_m))/2 < f(a_m)$. For $f'_+(w; w - a_m) < 0$ the result follows by taking $z_m := w$ and so we assume that $f'_+(w; w - a_m) \geq 0$. If $y_\lambda := w + \lambda(y - w)$ and the function $\theta : [0,1] \longrightarrow \mathbf{R}$ is given by $\theta(\lambda) := f(y_\lambda) + \lambda(f(w) - f(y))$ then clearly $\theta(0) = \theta(1) = f(w)$. Moreover, computing $\theta'_+(0; 1)$ yields

$$\begin{aligned}
\theta'_+(0;1) &= \limsup_{t \downarrow 0} \frac{\theta(0+t) - \theta(0)}{t} \\
&= \limsup_{t \downarrow 0} \frac{f(w + t(y-w)) - f(w) + t(f(w) - f(y))}{t} \\
&= f'_+(w; y - w) + f(w) - f(y).
\end{aligned}$$

Since by the definition of w it is easy to verify (remember $0 < \eta < 1$) that $y - w = (1-\eta)/\eta(w - a_m)$ we obtain by Lemma 5.2.1 that

$$f'_+(w; y - w) = \tfrac{1-\eta}{\eta} f'_+(w; w - a_m) \geq 0$$

and this yields using $f(w) > f(y)$ that $\theta'_+(0;1) > 0$. Hence, there exists some $\varepsilon > 0$ satisfying $\theta(\varepsilon) > \theta(0) = \theta(1)$ and since by the continuity of f the function θ is also continuous one can find some $0 < \xi < 1$ with $\theta(\xi) = \max_{0 \leq \lambda \leq 1} \theta(\lambda)$. This implies that $\theta'_+(\xi; \delta) \leq 0$ for each $\delta \in \{-1, 1\}$, and hence in particular $\theta'_+(\xi; 1) \leq 0$. Computing this derivative yields

$$\begin{aligned}
0 \geq \theta'_+(\xi;1) &= \limsup_{t \downarrow 0} \frac{\theta(\xi+t) - \theta(\xi)}{t} \\
&= \limsup_{t \downarrow 0} \frac{f(y_\xi + t(y-w)) - f(y_\xi) + t(f(w) - f(y))}{t} \\
&= f'_+(y_\xi; y - w) + f(w) - f(y)
\end{aligned}$$

and so $f'_+(y_\xi; y - w) \leq f(y) - f(w) < 0$. Since $w = \eta y + (1-\eta)a_m$ for some $0 < \eta < 1$ we obtain that $y_\xi - a_m = (\eta + \xi(1-\eta))/(1-\eta)(y - w)$ with $(\eta + \xi(1-\eta))/(1-\eta) > 0$ and hence by the previous inequality and Lemma 5.2.1 it follows that

$f'_+(y_\xi; y_\xi - a_m) < 0$. Observe now by the convexity of $\mathcal{L}_f^<(f(a_m)) \cap \mathcal{L}_{\bar{g}}^{\leq}(0) \cap \mathcal{L}_{\bar{n}}^{\leq}(r)$ and $y, w \in \mathcal{L}_f^<(f(a_m)) \cap \mathcal{L}_{\bar{g}}^{\leq}(0) \cap \mathcal{L}_{\bar{n}}^{\leq}(r)$ that y_ξ belongs to the same set and so the result is proven by taking $z_m := y_\xi$. □

An immediate corollary of this result is the following.

Corollary 5.4.1. *The point a_m is an optimal solution of problem (Q) if and only if every element z of the set $\mathcal{L}_f^<(f(a_m)) \cap \mathcal{L}_{\bar{g}}^{\leq}(0) \cap \mathcal{L}_{\bar{n}}^{\leq}(r)$ satisfies $f'_+(z; z - a_m) \geq 0$.*

Proof: If a_m is an optimal solution of (Q) then $\mathcal{L}_f^<(f(a_m)) \cap \mathcal{L}_{\bar{g}}^{\leq}(0) \cap \mathcal{L}_{\bar{n}}^{\leq}(r)$ is clearly empty and so the above implication is trivially satisfied. Suppose now that a_m is not optimal. In this case it follows by Assumption 3.1.1 that the set $\mathcal{L}_f^<(f(a_m)) \cap \mathcal{L}_{\bar{g}}^{\leq}(0) \cap \mathcal{L}_{\bar{n}}^{\leq}(r)$ is nonempty and by Lemma 5.4.1 this implies that at least one element z of the set $\mathcal{L}_f^<(f(a_m)) \cap \mathcal{L}_{\bar{g}}^{\leq}(0) \cap \mathcal{L}_{\bar{n}}^{\leq}(r)$ satisfies $f'_+(z; z - a_m) < 0$. This proves the result. □

Notice that Corollary 5.4.1 provides an optimality check. As a result of this, we observe if a_m is a nonoptimal "bad" point, i.e. $a_m \in \Gamma_f$, that there always exists a "good" point z_m. Moreover, Lemma 5.4.1 also gives us information how to find such a "good" point after the detection by some global search method of a point y belonging to $\mathcal{L}_f^<(f(a_m)) \cap \mathcal{L}_{\bar{g}}^{\leq}(0) \cap \mathcal{L}_{\bar{n}}^{\leq}(r)$. The previous observations lead to Algorithm 5.1, describing the application of an objective cut (replacing Algorithm 3.7) in Algorithm 3.11.

Observe, since this algorithm always starts by testing whether $a_m \in \Gamma_f$, that it is suitable for instances (like the functions discussed in Section 5.3) where this test can be efficiently implemented. On the other hand, instances where this test is computational expensive or even not implementable one may prefer to apply Algorithm 5.2. The implementation of these algorithms to generate an objective cut depends on the implementation of the procedures marked as framed statements. The procedure needed in Algorithm 5.1 to decide whether a_m belongs to Γ_f or not amounts in general to solving (S) and is highly problem specific. Hence it is not possible to describe it for the general case. Observe that Section 5.3 exhibits computable instances. We also refer to Chapter 6 for other examples. Similar comments apply to the procedures used to obtain elements from the normal cones.

if $f(\boldsymbol{a}_m) \leq \ell_{m-1}$ *then* let $\ell_m := f(\boldsymbol{a}_m)$ and $\boldsymbol{a}'_m := \boldsymbol{a}_m$
else let $\ell_m := \ell_{m-1}$ and $\boldsymbol{a}'_m := \boldsymbol{a}'_{m-1}$;
if $\boxed{\boldsymbol{a}_m \notin \Gamma_f}$
then $\boxed{\text{select a nonzero } \boldsymbol{a}^\star_m \in \mathcal{N}_{\mathcal{L}^{<}_f}(\boldsymbol{a}_m)}$ and let $\alpha_m := 0$
else *if* $\ell_m < f(\boldsymbol{a}_m)$
then $\boxed{\text{search }]\boldsymbol{a}'_{m-1}, \boldsymbol{a}_m[\text{ for a } \boldsymbol{z}_m \text{ satisfying } f'_+(\boldsymbol{z}_m; \boldsymbol{z}_m - \boldsymbol{a}_m) < 0}$;
if $f(\boldsymbol{z}_m) \leq \ell_m$ *then* redefine $\ell_m := f(\boldsymbol{z}_m)$ and $\boldsymbol{a}'_m := \boldsymbol{z}_m$;
else let $\mathcal{X}_m := \mathcal{L}^{<}_f(f(\boldsymbol{a}_m)) \cap \mathcal{L}^{\leq}_g(0) \cap \mathcal{L}^{\leq}_n(r)$;
$\boxed{\text{search } \mathcal{X}_m \text{ for a } \boldsymbol{z}_m \text{ satisfying } f'_+(\boldsymbol{z}_m; \boldsymbol{z}_m - \boldsymbol{a}_m) < 0}$;
if $\boldsymbol{z}_m$ is not found *then* stop, $\boldsymbol{a}_m$ is optimal
else redefine $\ell_m := f(\boldsymbol{z}_m)$ and $\boldsymbol{a}'_m := \boldsymbol{z}_m$;
$\boxed{\text{select a nonzero } \boldsymbol{z}^\star_m \in \mathcal{N}_{\mathcal{L}^{<}_f}(\boldsymbol{z}_m)}$;
let $\alpha_m := (\boldsymbol{z}^{\star\top}_m(\boldsymbol{a}_m - \boldsymbol{z}_m))/\sqrt{\boldsymbol{z}^{\star\top}_m \boldsymbol{A}_m \boldsymbol{z}^\star_m}$;
let $\delta_m := \frac{s^2(1-\alpha_m^2)}{s^2-1}$, $\sigma_m := \frac{2(1+s\alpha_m)}{(s+1)(1+\alpha_m)}$ and $\tau_m := \frac{1+s\alpha_m}{s+1}$

Algorithm 5.1: Application of an objective cut in the case of a "cheap" oracle

The remaining two procedures belong in general to the field of global optimization and the derivation of efficient algorithms to implement them falls beyond the scope of this book.

In the end of Subsection 5.6.2 we discuss a possible algorithm. Although this is based on the inefficient grid search method (Ref. [107]) it is useful as an example.

Regardless of the search method used, the complicated nature of $\mathcal{X}_m$ makes it much more convenient to apply such a search on an easier set containing $\mathcal{X}_m$. Observe that Section 5.6 discusses the construction of an easy *box* or *hyperrectangle* containing this set. In this case each candidate point $\boldsymbol{z}_m$ belonging to this easier set is first tested for membership of $\mathcal{X}_m$ before being checked for the satisfaction of $f'_+(\boldsymbol{z}_m; \boldsymbol{z}_m - \boldsymbol{a}_m) < 0$.

if $f(\boldsymbol{a}_m) \leq \ell_{m-1}$ *then* let $\ell_m := f(\boldsymbol{a}_m)$ and $\boldsymbol{a}'_m := \boldsymbol{a}_m$
else let $\ell_m := \ell_{m-1}$ and $\boldsymbol{a}'_m := \boldsymbol{a}'_{m-1}$;
if $\ell_m < f(\boldsymbol{a}_m)$
then search $]\boldsymbol{a}'_{m-1}, \boldsymbol{a}_m[$ for a $\boldsymbol{z}_m$ satisfying $f'_+(\boldsymbol{z}_m; \boldsymbol{z}_m - \boldsymbol{a}_m) < 0$;
if $f(\boldsymbol{z}_m) \leq \ell_m$ *then* redefine $\ell_m := f(\boldsymbol{z}_m)$ and $\boldsymbol{a}'_m := \boldsymbol{z}_m$;
else let $\mathcal{X}_m := \mathcal{L}^{<}_{f}(f(\boldsymbol{a}_m)) \cap \mathcal{L}^{\leq}_{\bar{g}}(0) \cap \mathcal{L}^{\leq}_{\bar{n}}(r)$;
search $\mathcal{X}_m$ for a $\boldsymbol{z}_m$ satisfying $f'_+(\boldsymbol{z}_m; \boldsymbol{z}_m - \boldsymbol{a}_m) < 0$;
if $\boldsymbol{z}_m$ is not found *then* stop, $\boldsymbol{a}_m$ is optimal
else redefine $\ell_m := f(\boldsymbol{z}_m)$ and $\boldsymbol{a}'_m := \boldsymbol{z}_m$;
select a nonzero $\boldsymbol{z}^\star_m \in \mathcal{N}_{\mathcal{L}^{<}_{f}}(\boldsymbol{z}_m)$;
let $\alpha_m := (\boldsymbol{z}^{\star\mathsf{T}}_m(\boldsymbol{a}_m - \boldsymbol{z}_m)) / \sqrt{\boldsymbol{z}^{\star\mathsf{T}}_m \boldsymbol{A}_m \boldsymbol{z}^\star_m}$;
let $\delta_m := \frac{s^2(1-\alpha_m^2)}{s^2-1}$, $\sigma_m := \frac{2(1+s\alpha_m)}{(s+1)(1+\alpha_m)}$ and $\tau_m := \frac{1+s\alpha_m}{s+1}$

Algorithm 5.2: Application of an objective cut in the case of an “expensive” oracle

To justify the complicated structure of these algorithms we observe that after the value ℓ_m has been found for the first time, say in iteration $k < m$, and the corresponding point $\boldsymbol{a}'_m$ has been memorized, it may happen that $f(\boldsymbol{a}_m) = \ell_m$. This may lead to $\boldsymbol{a}'_m$ being “cut away” in the next ellipsoid, see Figure 5.7.

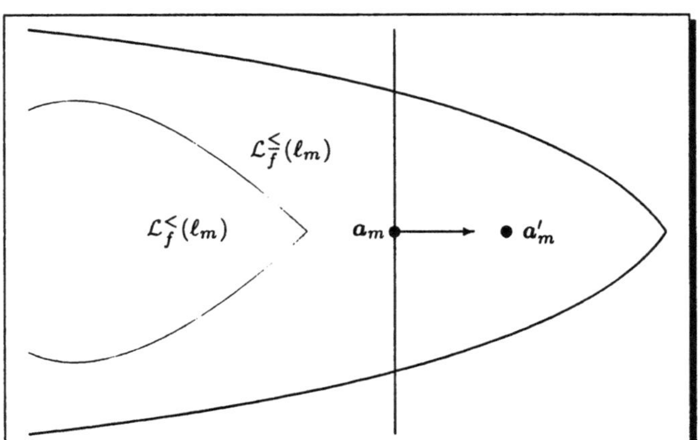

Figure 5.7: The best point found so far may be cut away

Hence we should always keep the last point yielding value ℓ_m. However, we have to be extremely careful in the way we define "last". Observe that while searching for z_m in $]a'_{m-1}, a_m[$ it may happen that $f(z_m) \leq \ell_m$ and hence z_m itself should be considered as a potential best recorded feasible solution.

For an illustration see Figure 5.8 where the first picture shows the level sets and the second picture a slice of the graph of f.

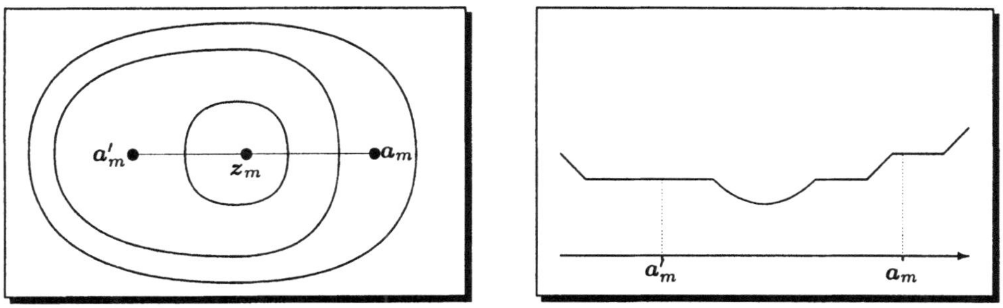

Figure 5.8: A line search may improve ℓ_m

On the other hand, if z_m is found in $\mathcal{X}_m$ this point is inside $\mathcal{L}_f^{<}(f(a_m))$ and hence $f(z_m) < f(a_m)$. Observe that $f(a_m)$ equals ℓ_m in this case. Hence we have just justified the reason for also considering z_m in order to update ℓ_m and a'_m.

With respect to Algorithm 5.1 and Algorithm 5.2 we like to point out that once a point z_m is found satisfying $f'_+(z_m; z_m - a_m) < 0$ one "only" needs to find an element of $\partial h_-(z_m - a_m)$ with $h_-(\cdot)$ the restriction of $f'_+(z_m;\cdot)$ to cl $\left(\mathcal{D}_f^{<}(z_m)\right)$. By the remarks after Lemma 5.2.7, one may often expect $\nabla h_-(z_m - a_m)$ to exist and hence obtaining an element of $\partial h_-(z_m - a_m)$ may be easier. Finally we like to stress that Algorithm 5.1 and Algorithm 5.2 must be seen as black boxes in the sense that a'_k and ℓ_k should be regarded in the following discussion as the values generated at the end of each iteration k. It now remains to analyze the properties of Algorithm 3.11 when Algorithm 5.1 or Algorithm 5.2 replace Algorithm 3.7 in the implementation of the objective cuts. In order to do so we consider the sequences ℓ_k and a'_k for $k \geq 0$. It is only relevant to look at these sequences for finite values of ℓ_k and so we consider $\underline{k} := \inf\{k \geq 0 : \ell_k < \infty\}$. Clearly at iteration k with $k < \underline{k}$ either a norm or a constraint cut is applied and similarly as for the convex case it is not difficult to verify that[2]

$$\mathcal{L}_f^<(\ell_k) \cap \mathcal{L}_g^\leq(0) \cap \mathcal{L}_n^\leq(r) \subseteq \mathcal{E}(\boldsymbol{A}_k; \boldsymbol{a}_k) \cap \mathcal{L}_{h_k}^\leq(\beta_k) \subseteq \mathcal{E}(\boldsymbol{A}_{k+1}; \boldsymbol{a}_{k+1}). \qquad (5.14)$$

By Assumption 3.1.1 and the above inclusion it follows from a simple volume argument that $\underline{k}$ is finite. We now consider iteration $k \geq \underline{k}$ and introduce the following definition.

Definition 5.4.1. *Algorithm 3.11 with the implementation of an objective cut done by Algorithm 5.1 or Algorithm 5.2 is said to be consistent at iteration $k \geq \underline{k}$ if*

1. *the cut generated is a valid cut of $\mathcal{E}(\boldsymbol{A}_k; \boldsymbol{a}_k)$ with $0 \leq \alpha_k \leq 1$,*
2. *(5.14) holds and*
3. *$\boldsymbol{a}'_k \in \mathcal{E}(\boldsymbol{A}_{k+1}; \boldsymbol{a}_{k+1})$.*

We now prove the consistency of Algorithm 3.11 for quasiconvex functions.

Lemma 5.4.2. *Algorithm 3.11 in combination with Algorithm 5.1 or Algorithm 5.2 is consistent at iteration $\underline{k}$.*

Proof: From the definition of $\underline{k}$ it follows that $\boldsymbol{a}_{\underline{k}}$ is the first center generated to be feasible and to satisfy $n(\boldsymbol{a}_{\underline{k}}) \leq r$. Hence Algorithm 5.1 or Algorithm 5.2 is executed for the first time and this gives rise to two possible situations. The first, only possible if Algorithm 5.1 is used, corresponds to $\boldsymbol{a}_{\underline{k}} \notin \Gamma_f$ and hence a separation oracle as discussed in Subsection 5.2.2 can be applied to $\boldsymbol{a}_{\underline{k}}$ yielding some point $\boldsymbol{a}^\star_{\underline{k}} \in \mathcal{N}_{\mathcal{L}_f^<}(\boldsymbol{a}_{\underline{k}})$. It follows from Algorithm 5.1 that the corresponding cut $\mathcal{L}_{h_{\underline{k}}}^=(0)$ with $h_{\underline{k}}(\boldsymbol{x}) := \boldsymbol{a}^{\star\top}_{\underline{k}}(\boldsymbol{x} - \boldsymbol{a}_{\underline{k}})$ is a valid central cut due to $\boldsymbol{a}_{\underline{k}} \in \mathcal{E}(\boldsymbol{A}_{\underline{k}}; \boldsymbol{a}_{\underline{k}})$ and $\boldsymbol{a}_{\underline{k}} \in \mathcal{L}_{h_{\underline{k}}}^=(0)$. Moreover, since $\boldsymbol{a}'_{\underline{k}} = \boldsymbol{a}_{\underline{k}}$ and $\ell_{\underline{k}} = f(\boldsymbol{a}_{\underline{k}})$ we obtain that $\boldsymbol{a}'_{\underline{k}} \in \mathcal{E}(\boldsymbol{A}_{\underline{k}}; \boldsymbol{a}_{\underline{k}}) \cap \mathcal{L}_{h_{\underline{k}}}^\leq(0) \subseteq \mathcal{E}(\boldsymbol{A}_{\underline{k}+1}; \boldsymbol{a}_{\underline{k}+1})$. Moreover, we have $\mathcal{L}_f^<(\ell_{\underline{k}}) \subseteq \mathcal{L}_{h_{\underline{k}}}^\leq(0)$ and from (5.14) with $k = \underline{k} - 1$ it follows that

$$\mathcal{L}_f^<(\ell_{\underline{k}}) \cap \mathcal{L}_g^\leq(0) \cap \mathcal{L}_n^\leq(r) \subseteq \mathcal{E}(\boldsymbol{A}_{\underline{k}}; \boldsymbol{a}_{\underline{k}}) \cap \mathcal{L}_{h_{\underline{k}}}^\leq(\beta_{\underline{k}}) \subseteq \mathcal{E}(\boldsymbol{A}_{\underline{k}+1}; \boldsymbol{a}_{\underline{k}+1}).$$

[2] Observe for $k < \underline{k}$ that $\mathcal{L}_f^<(\ell_k) = \mathcal{L}_f^<(\infty) = \mathbf{R}^s$.

Hence, the result is verified if $a_{\underline{k}}$ is a "good" point. On the other hand it may happen that Algorithm 5.1 searches $\mathcal{X}_{\underline{k}}$ for some $z_{\underline{k}}$ satisfying $f'_+(z_{\underline{k}}; z_{\underline{k}} - a_{\underline{k}}) < 0$ because $a_{\underline{k}}$ is a "bad" point. This is the strategy followed by Algorithm 5.2. If such a point $z_{\underline{k}}$ is found (otherwise the algorithm stops by Corollary 5.4.1) it follows from the definition of $\mathcal{X}_{\underline{k}}$ that $f(z_{\underline{k}}) < f(a_{\underline{k}})$. Moreover, the separation oracle discussed in Subsection 5.2.2 applied to $z_{\underline{k}}$ yields some point $z^\star_{\underline{k}} \in \mathcal{N}_{\mathcal{L}^<_f}(z_{\underline{k}})$ and this originates the cut $\mathcal{L}^=_{\overline{h}_{\underline{k}}}(\beta_{\underline{k}})$ with $h_{\underline{k}}(x) := z^{\star\mathsf{T}}_{\underline{k}}(x - a_{\underline{k}})$ and $\beta_{\underline{k}} := z^{\star\mathsf{T}}_{\underline{k}}(z_{\underline{k}} - a_{\underline{k}})$. In order to show that this is a valid deep cut we first observe by (5.14) with $k = \underline{k} - 1$ that $z_{\underline{k}} \in \mathcal{X}_{\underline{k}} \subseteq \mathcal{L}^{\leq}_{\overline{g}}(0) \cap \mathcal{L}^{\leq}_{\overline{n}}(r) \subseteq \mathcal{E}(A_{\underline{k}}; a_{\underline{k}})$. Since $z_{\underline{k}} \in \mathcal{L}^=_{\overline{h}_{\underline{k}}}(\beta_{\underline{k}})$ this cut is valid. Moreover, by Lemma 5.2.7 it follows that $\lambda z^\star_{\underline{k}} \in \partial h_-(0)$ for some $\lambda > 0$ and this implies by Lemma 5.4.1 and Lemma 5.2.7 that

$$
\begin{aligned}
0 &> f'_+(z_{\underline{k}}; z_{\underline{k}} - a_{\underline{k}}) = \sup\{x^{\star\mathsf{T}}(z_{\underline{k}} - a_{\underline{k}}) : x^\star \in \partial h_-(0)\} \\
&\geq \lambda z^{\star\mathsf{T}}_{\underline{k}}(z_{\underline{k}} - a_{\underline{k}}) = \lambda \beta_{\underline{k}}.
\end{aligned}
$$

Hence $\beta_{\underline{k}} < 0$ and the cut is deep. It follows for this case using $z^\star_{\underline{k}} \in \mathcal{N}_{\mathcal{L}^<_f}(z_{\underline{k}})$ and $a'_{\underline{k}} = z_{\underline{k}}$ that $a'_{\underline{k}} \in \mathcal{E}(A_{\underline{k}}; a_{\underline{k}}) \cap \mathcal{L}^{\leq}_{\overline{h}_{\underline{k}}}(\beta_{\underline{k}}) \subseteq \mathcal{E}(A_{\underline{k}+1}; a_{\underline{k}+1})$. Finally, since $\mathcal{L}^<_f(\ell_{\underline{k}}) = \mathcal{L}^<_f(f(a'_{\underline{k}})) \subseteq \mathcal{L}^{\leq}_{\overline{h}_{\underline{k}}}(\beta_{\underline{k}})$ we obtain from (5.14) that

$$
\mathcal{L}^<_f(\ell_{\underline{k}}) \cap \mathcal{L}^{\leq}_{\overline{g}}(0) \cap \mathcal{L}^{\leq}_{\overline{n}}(r) \subseteq \mathcal{E}(A_{\underline{k}}; a_{\underline{k}}) \cap \mathcal{L}^{\leq}_{\overline{h}_{\underline{k}}}(\beta_{\underline{k}}) \subseteq \mathcal{E}(A_{\underline{k}+1}; a_{\underline{k}+1})
$$

and so the desired result is proven in both cases. □

The next result needs to be proven in order to complete our analysis of the algorithm.

Lemma 5.4.3. *Algorithm 3.11 in combination with Algorithm 5.1 or Algorithm 5.2 is consistent at iteration $k > \underline{k}$ if it is consistent at iteration $k - 1$.*

Proof: If a norm or a constraint cut is applied at iteration k then, just as for the convex case, it follows that $\mathcal{L}^=_{\overline{h}_k}(\beta_k)$ is a valid deep cut. Moreover, $\ell_k = \ell_{k-1}$ and either $\mathcal{L}^{\leq}_{\overline{g}}(0) \subseteq \mathcal{L}^{\leq}_{\overline{h}_k}(\beta_k)$ (constraint cut) or $\mathcal{L}^{\leq}_{\overline{n}}(r) \subseteq \mathcal{L}^{\leq}_{\overline{h}_k}(\beta_k)$ (norm cut) and hence it follows by our assumption of consistency at iteration $k - 1$ that

$$
\mathcal{L}^<_f(\ell_k) \cap \mathcal{L}^{\leq}_{\overline{g}}(0) \cap \mathcal{L}^{\leq}_{\overline{n}}(r) \subseteq \mathcal{E}(A_k; a_k) \cap \mathcal{L}^{\leq}_{\overline{h}_k}(\beta_k) \subseteq \mathcal{E}(A_{k+1}; a_{k+1}).
$$

Furthermore, since $a'_k = a'_{k-1}$ and $a'_k \in \mathcal{L}^{\leq}_{\bar{g}}(0) \cap \mathcal{L}^{\leq}_{\bar{n}}(r) \subseteq \mathcal{L}^{\leq}_{\bar{h}_k}(\beta_k)$ we obtain again by our assumption that $a'_k \in \mathcal{E}(A_k; a_k) \cap \mathcal{L}^{\leq}_{\bar{h}_k}(\beta_k) \subseteq \mathcal{E}(A_{k+1}; a_{k+1})$ and this proves the result for a norm or a constraint cut. On the other hand, if an objective cut is applied at iteration k and the point a_k is a "good" point, i.e. $a_k \notin \Gamma_f$, it follows that $a_k \in \mathrm{cl}\left(\mathcal{L}^{<}_f(f(a_k))\right)$ and from Algorithm 5.1 that the constructed cut $\mathcal{L}^{=}_{\bar{h}_k}(0)$ is a valid (from $a_k \in \mathcal{E}(A_k; a_k)$) central cut. Moreover, if $\ell_{k-1} \geq f(a_k)$ then $a'_k = a_k$ and so $a'_k \in \mathcal{L}^{\leq}_{\bar{h}_k}(0)$. Hence, our assumption yields in any case that $a'_k \in \mathcal{E}(A_k; a_k) \cap \mathcal{L}^{\leq}_{\bar{h}_k}(\beta_k) \subseteq \mathcal{E}(A_{k+1}; a_{k+1})$. Similarly, since $\ell_k = f(a_k) \leq \ell_{k-1}$ and $a'_k \in \mathcal{N}_{\mathcal{L}^{<}_f}(a_k)$ it follows that $\mathcal{L}^{<}_f(\ell_k) = \mathcal{L}^{<}_f(f(a_k) \subseteq \mathcal{L}^{\leq}_{\bar{h}_k}(0)$ and

$$\mathcal{L}^{<}_f(\ell_k) \cap \mathcal{L}^{\leq}_{\bar{g}}(0) \cap \mathcal{L}^{\leq}_{\bar{n}}(r) \subseteq \mathcal{E}(A_k; a_k) \cap \mathcal{L}^{\leq}_{\bar{h}_k}(\beta_k) \subseteq \mathcal{E}(A_{k+1}; a_{k+1})$$

which covers the case of an objective cut performed with a "good" a_k. Now it remains to analyze the situation when Algorithm 5.1 or Algorithm 5.2 must search for a z_k satisfying $f'_+(z_k; z_k - a_k) < 0$. We first consider the case when the interval $]a'_{k-1}, a_k[$ is searched. From our assumption we know that $a'_{k-1} \in \mathcal{E}(A_k; a_k)$ and so by the convexity of $\mathcal{E}(A_k; a_k)$ it follows that $]a'_{k-1}, a_k[\subseteq \mathcal{E}(A_k; a_k)$ and hence $z_k \in \mathcal{E}(A_k; a_k)$. Moreover, the cut $\mathcal{L}^{=}_{\bar{h}_k}(\beta_k)$ with $h_k(x) := z_k^{\star\mathsf{T}}(x - a_k)$ and $\beta_k := z_k^{\star\mathsf{T}}(z_k - a_k)$ satisfies $z_k \in \mathcal{L}^{=}_{\bar{h}_k}(\beta_k)$ and so it is valid. In order to show that this is a deep cut we observe by Lemma 5.2.7 that $\lambda z_k^\star \in \partial h_-(0)$ for some $\lambda > 0$ and this implies by Lemma 5.4.1 and Lemma 5.2.7 that

$$\begin{aligned} 0 &> f'_+(z_k; z_k - a_k) = \sup\{x^{\star\mathsf{T}}(z_k - a_k) : x^\star \in \partial h_-(0)\} \\ &\geq \lambda z_k^{\star\mathsf{T}}(z_k - a_k) = \lambda\beta_k. \end{aligned}$$

Hence $\beta_k < 0$ and the cut is deep. It follows from Algorithm 5.1 or Algorithm 5.2 that if $a'_k = a'_{k-1}$ then $a'_k \in \mathcal{L}^{<}_f(f(z_k)) \subseteq \mathcal{L}^{\leq}_{\bar{h}_k}(\beta_k)$ and by our assumption this yields $a'_k \in \mathcal{E}(A_k; a_k) \cap \mathcal{L}^{\leq}_{\bar{h}_k}(\beta_k) \subseteq \mathcal{E}(A_{k+1}; a_{k+1})$. On the other hand, if $a'_k = z_k$ then we clearly obtain due to $f'_+(z_k; z_k - a_k) < 0$ that $a'_k \in \mathrm{cl}\left(\mathcal{L}^{<}_f(f(z_k))\right) \subseteq \mathcal{L}^{\leq}_{\bar{h}_k}(\beta_k)$. Since in this case $f(z_k) < \ell_{k-1}$ this implies by our assumption that $a'_k = z_k \in \mathcal{L}^{<}_f(\ell_{k-1}) \cap \mathcal{L}^{\leq}_{\bar{g}}(0) \cap \mathcal{L}^{\leq}_{\bar{n}}(r) \subseteq \mathcal{E}(A_k; a_k)$ and so $a'_k \in \mathcal{E}(A_k; a_k) \cap \mathcal{L}^{\leq}_{\bar{h}_k}(\beta_k) \subseteq \mathcal{E}(A_{k+1}; a_{k+1})$. Also, $\ell_k \leq \min\{\ell_{k-1}, f(z_k)\}$ leads to $\mathcal{L}^{<}_f(\ell_k) \subseteq \mathcal{L}^{<}_f(f(z_k)) \subseteq \mathcal{L}^{\leq}_{\bar{h}_k}(\beta_k)$ and from this we conclude that

$$\mathcal{L}^{<}_f(\ell_k) \cap \mathcal{L}^{\leq}_{\bar{g}}(0) \cap \mathcal{L}^{\leq}_{\bar{n}}(r) \subseteq \mathcal{E}(A_k; a_k) \cap \mathcal{L}^{\leq}_{\bar{h}_k}(\beta_k) \subseteq \mathcal{E}(A_{k+1}; a_{k+1})$$

and the result is also proven for this case. At last we consider the case when Algorithm 5.1 or Algorithm 5.2 searches $\mathcal{X}_k$. The validity of the corresponding deep cut can be established similarly as for the previous case using $z_k \in \mathcal{X}_k \subseteq \mathcal{E}(A_k; a_k)$. Finally, since $a'_k = z_k$ we obtain that $a'_k \in \mathcal{E}(A_k; a_k) \cap \mathcal{L}^{\leq}_{\overline{h}_k}(\beta_k) \subseteq \mathcal{E}(A_{k+1}; a_{k+1})$ and since $\mathcal{L}^{<}_{f}(\ell_k) = \mathcal{L}^{<}_{f}(f(z_k)) \subseteq \mathcal{L}^{\leq}_{\overline{h}_k}(\beta_k)$ and $\ell_k \leq \ell_{k-1}$ it follows that

$$\mathcal{L}^{<}_{f}(\ell_k) \cap \mathcal{L}^{\leq}_{\overline{g}}(0) \cap \mathcal{L}^{\leq}_{\overline{n}}(r) \subseteq \mathcal{E}(A_k; a_k) \cap \mathcal{L}^{\leq}_{\overline{h}_k}(\beta_k) \subseteq \mathcal{E}(A_{k+1}; a_{k+1})$$

concluding the proof. □

The following result is an immediate consequence of Lemma 5.4.2 and Lemma 5.4.3.

Theorem 5.4.1. *If Assumption 3.1.1 holds then Algorithm 3.11 with the application of an objective cut carried out by Algorithm 5.1 or Algorithm 5.2 satisfies at every iteration k the inclusion*

$$\mathcal{L}^{<}_{f}(\ell_k) \cap \mathcal{L}^{\leq}_{\overline{n}}(r) \cap \mathcal{L}^{\leq}_{\overline{g}}(0) \subseteq \mathcal{E}(A_k; a_k) \cap \mathcal{L}^{\leq}_{\overline{h}_k}(\beta_k).$$

Proof: If the present iteration $k < \underline{k}$ then the desired result follows by (5.14) since $\ell_k = \infty$. On the other hand, Lemma 5.4.2 establishes the consistency at iteration $\underline{k}$ while repeated applications of Lemma 5.4.3 cover the subsequent iterations. □

Observe that in absence of other stopping criteria Algorithm 3.11 is only stopped at iteration m if Algorithm 5.1 (or Algorithm 5.2) cannot find a z_m satisfying $f'(z_m; z_m - a_m) < 0$. By Corollary 5.4.1 this certifies a_m as an optimal solution. In order to enable the implementation of this global search procedure Section 5.6 constructs an hyperrectangle containing $\mathcal{X}_m$. However, before that we give a proof of convergence of Algorithm 3.11 in the next section.

5.5. Convergence proof

In this section we present a rate of convergence proof for Algorithm 3.11 applying objective cuts by means of Algorithm 5.1 or Algorithm 5.2. The complexity of this proof depends much more on the presence or not of a constraint function than the case studied in Chapter 3. Therefore, we analyze separately the unconstrained and the constrained case.

Observe, if the algorithm stops at iteration m then $\boldsymbol{a}_m$ is optimal and so the convergence of the algorithm is established for this case. Therefore this section considers only the case for which the algorithm does not stop after a finite number of iterations and this means in particular that $\ell_m > f(\boldsymbol{x}_*)$. Recall that $a := (s^2 - 1)/(s^2)$ and $b := \sqrt[s]{(s+1)/(s-1)}$ and that these numbers satisfy $ab > 1$ and so $1/\sqrt{ab} < 1$.

Clearly, $0 \leq \alpha_m < 1$ since we have seen that $0 \leq \alpha_m \leq 1$ and if $\alpha_m = 1$ then the next ellipsoid would reduce to an optimal solution and hence the algorithm would stop in finitely many iterations.

5.5.1. The unconstrained quasiconvex program

For (Q) unconstrained we can establish the following rate of convergence result.

Theorem 5.5.1. *If (Q) is unconstrained, f is Lipschitz continuous on $\mathcal{L}_{\overline{n}}^{\leq}(r)$ with Lipschitz constant L_f and the algorithm does not stop in a finite number of iterations then there exists some m_0 such that*

$$0 \leq \ell_m - f(\boldsymbol{x}_*) \leq \frac{L_f r}{\sqrt[s]{2}} \left(\frac{1}{\sqrt{ab}}\right)^m \prod_{k=0}^{m-1} \sqrt{1-\alpha_k^2} \sqrt[2s]{\frac{1-\alpha_k}{1+\alpha_k}}.$$

for every $m \geq m_0$.

Proof: Since f is continuous it follows by Theorem 3.2.1 that $\ell_m \downarrow f(\boldsymbol{x}_*)$ and so by Assumption 3.1.1 there exists some m_0 such that

$$\boldsymbol{x}_* + \frac{\ell_m - f(\boldsymbol{x}_*)}{L_f} \in \mathcal{L}_{\overline{n}}^{\leq}(r)$$

for every $m \geq m_0$. Also by Lemma 3.2.5 we obtain that

$$\boldsymbol{x}_* + \frac{\ell_m - f(\boldsymbol{x}_*)}{L_f}\mathcal{B} \subseteq \mathcal{L}_f^{<}(\ell_m)$$

and these two conditions together with Theorem 5.4.1 yield that

$$x_* + \frac{\ell_m - f(x_*)}{L_f}\mathcal{B} \subseteq \mathcal{L}_f^{\leq}(\ell_m) \cap \mathcal{L}_{\bar{n}}^{\leq}(r) \subseteq \mathcal{E}(A_m; a_m) \cap \mathcal{L}_{\bar{h}_m}^{\leq}(\beta_m).$$

Thus,

$$\operatorname{vol}\left(x_* + \frac{\ell_m - f(x_*)}{L_f}\mathcal{B}\right) \leq \operatorname{vol}\left(\mathcal{E}(A_m; a_m) \cap \mathcal{L}_{\bar{h}_m}^{\leq}(\beta_m)\right) \leq \frac{1}{2}\operatorname{vol}(\mathcal{E}(A_m; a_m))$$

and since $\operatorname{vol}(\mathcal{E}(A_m; a_m)) = \sqrt{\det(A_m)}v_s$, with $\det(A_m)$ given in (3.8) and $v_s :=$ $\operatorname{vol}(\mathcal{B})$, computing these volumes gives

$$\left(\frac{\ell_m - f(x_*)}{L_f}\right)^s v_s \leq \frac{1}{2}\sqrt{r^{2s}\prod_{k=0}^{m-1}\left(\frac{1}{ab}\right)^s (1-\alpha_k^2)^s \left(\frac{1-\alpha_k}{1+\alpha_k}\right)} v_s.$$

Dividing by v_s, raising both sides to s^{-1} and multiplying by L_f yields the desired result. □

In the next section we give a rate of convergence proof for the constrained program.

5.5.2. The constrained quasiconvex program

The analysis of this case is preceded by Subsection 3.3.1, which does not use the convexity of f.

By Theorem 3.3.1 and Lemma 3.3.3 it follows that there exists some m_0 such that $\omega_m = \min\{\omega_m^g, \omega_m^f\}$ for $m \geq m_0$. This means that for m big enough we only have to study the behavior of the sequences ω_m^g and ω_m^f.

A useful result for the proof of the main theorem is given by the next lemma which is an immediate consequence of Theorem 5.4.1.

Lemma 5.5.1. *If $\ell_{m_j} > f(x_*) + j\omega_{m_j}$ for some $m_j \in \mathbb{N}$ then $\mathcal{L}_f^{\leq}(f(x_*) + j\omega_{m_j}) \cap \mathcal{L}_g^{\leq}(0) \cap \mathcal{L}_{\bar{n}}^{\leq}(r) \subseteq \mathcal{E}(A_k; a_k)$ for every $k \leq m_j$.*

Proof: It follows from $k \leq m_j$ that $\ell_{k-1} \geq \ell_{m_j} > f(x_*) + j\omega_{m_j}$ and this implies that $\mathcal{L}_f^{\leq}(f(x_*) + j\omega_{m_j}) \subseteq \mathcal{L}_f^{<}(\ell_{k-1})$. The result follows now from Theorem 5.4.1. □

Notice that in order to prove the main convergence rate theorem and link the behavior of ω_m to ω_m^f we also need Assumption 3.2.1. Moreover, the next result is proven only for Lipschitz continuous quasiconvex functions. Recall that such a function is boundedly lower subdifferentiable (Definition 2.4.3).

Theorem 5.5.2. *If the algorithm does not stop in a finite number of iterations, f is Lipschitz continuous with Lipschitz constant L_f on $\mathbf{R}^s$, Assumption 3.1.1 and Assumption 3.2.1 hold then $\ell_m \downarrow f(x_*)$ with at least the same rate as $\omega_m \downarrow 0$.*

Proof: Clearly, by Theorem 3.2.1 the sequence ℓ_m converges to $f(x_*)$. If an optimal solution x_* exists satisfying $g(x_*) < 0$ and $n(x_*) < r$ then the proof of Theorem 5.5.1 applies. To start the analysis of the other case observe that $\limsup_{m\uparrow\infty} \frac{l_m - f(x_*)}{\omega_m} < \infty$ implies that the sequence $\ell_m - f(x_*)$ has at least the same convergence rate as the sequence ω_m. Suppose therefore that all the optimal solutions with $n(x_*) < r$ satisfy $g(x_*) = 0$ and assume by contradiction that

$$\limsup_{m\uparrow\infty} \frac{l_m - f(x_*)}{\omega_m} = \infty.$$

This implies that there exists an increasing sequence $\{m_j\}_{j=1}^{\infty}$ with $m_j \uparrow \infty$ and $\ell_{m_j} - f(x_*) > j\omega_{m_j}$. By Assumption 3.2.1 it follows that $g(\widehat{x}) := \min\{g(x) : x \in \mathcal{L}_n^{\leq}(r)\} = -\delta$ for some $\delta > 0$ and so $\widehat{x} \in \mathcal{L}_n^{\leq}(r)$ cannot be optimal. Hence, by Theorem 3.3.1 there exists some j_0 such that $\widehat{x} \notin \mathcal{L}_f^{\leq}(f(x_*) + j\omega_{m_j}) \cap \mathcal{L}_g^{\leq}(0) \cap \mathcal{L}_n^{\leq}(r)$ for every $j \geq j_0$. Since f is continuous it follows for every $j \geq j_0$ that there exists some $\lambda_j \in]0,1[$ such that $x_j := \lambda_j \widehat{x} + (1-\lambda_j)x_* \in \mathcal{L}_n^{\leq}(r)$ with $f(x_j) = f(x_*) + \frac{j}{2}\omega_{m_j}$. Moreover, by the convexity of g and the definition of $\widehat{x}$ we obtain

$$\frac{g(x_*) - g(x_j)}{\|x_* - x_j\|_2} \geq \frac{g(x_*) - g(\widehat{x})}{\|x_* - \widehat{x}\|_2}$$

and this yields using $g(x_*) = 0$ and $g(\widehat{x}) = -\delta$ that

$$-g(x_j) \geq \delta \frac{\|x_* - x_j\|_2}{\|x_* - \widehat{x}\|_2}. \tag{5.15}$$

By the Lipschitz continuity of f with Lipschitz constant L_f it follows that $\frac{j}{2}\omega_{m_j} = f(\boldsymbol{x}_j) - f(\boldsymbol{x}_*) \leq L_f\|\boldsymbol{x}_j - \boldsymbol{x}_*\|_2$ and so by (5.15)

$$-g(\boldsymbol{x}_j) \geq \frac{j\delta\omega_{m_j}}{2\|\boldsymbol{x}_* - \widehat{\boldsymbol{x}}\|_2 L_f}. \tag{5.16}$$

Consider now some arbitrary $k \leq m_j$ and suppose at iteration k a constraint cut is performed. By (5.16), the definition of $\boldsymbol{x}_j$ and Lemma 5.5.1 we obtain $\boldsymbol{x}_j \in \mathcal{L}_f^{\leq}(f(\boldsymbol{x}_*) + j\omega_{m_j}) \cap \mathcal{L}_g^{\leq}(0) \cap \mathcal{L}_n^{\leq}(r) \subseteq \mathcal{E}(\boldsymbol{A}_k; \boldsymbol{a}_k)$, $k \leq m_j$. Now it follows by the subgradient inequality and (3.3) that

$$g(\boldsymbol{x}_j) \geq g(\boldsymbol{a}_k) + \boldsymbol{a}_j^{\star\top}(\boldsymbol{x}_j - \boldsymbol{a}_k) \geq g(\boldsymbol{a}_k) - \sqrt{\boldsymbol{a}_j^{\star\top}\boldsymbol{A}_k\boldsymbol{a}_j^{\star}}$$

and so by Algorithm 3.8

$$\begin{aligned}(1-\alpha_k)\varepsilon_k\|\boldsymbol{a}_j^{\star}\|_2 &= \left(1 - \frac{g(\boldsymbol{a}_k)}{\sqrt{\boldsymbol{a}_j^{\star\top}\boldsymbol{A}_k\boldsymbol{a}_j^{\star}}}\right)\sqrt{\boldsymbol{a}_j^{\star\top}\boldsymbol{A}_k\boldsymbol{a}_j^{\star}} \\ &= \sqrt{\boldsymbol{a}_j^{\star\top}\boldsymbol{A}_k\boldsymbol{a}_j^{\star}} - g(\boldsymbol{a}_k) \geq -g(\boldsymbol{x}_j).\end{aligned}$$

This implies using $\|\boldsymbol{a}_j^{\star}\|_2 \leq L_g$ with L_g the Lipschitz constant of g on $\mathcal{L}_n^{\leq}(r)$ that

$$L_g\omega_{m_j}^g \geq -g(\boldsymbol{x}_j)$$

and hence by (5.16)

$$L_g\omega_{m_j}^g \geq \frac{j\delta\omega_{m_j}}{2\|\boldsymbol{x}_* - \widehat{\boldsymbol{x}}\|_2 L_f}. \tag{5.17}$$

On the other hand, if at iteration m_j an objective cut is performed we obtain by (3.15) and $\ell_{m_j} - f(\boldsymbol{x}_*) > j\omega_{m_j}$ that[3]

$$L_f\omega_{m_j}^f \geq \ell_{m_j} - f(\boldsymbol{x}_*) > j\omega_{m_j}. \tag{5.18}$$

Combining now (5.18) and (5.17) it follows for j sufficiently large and satisfying

$$\min\left\{\frac{j\delta}{2\|\boldsymbol{x}_* - \widehat{\boldsymbol{x}}\|_2 L_f L_g}, \frac{j}{L_f}\right\} > 1$$

[3]Observe that the derivation of a stopping rule in Subsection 3.3 holds for lower subdifferentiable functions.

that

$$\omega_{m_j} = \min\{\omega^g_{m_j}, \omega^f_{m_j}\} \geq \min\left\{\frac{j\delta}{2\|\boldsymbol{x}_* - \widehat{\boldsymbol{x}}\|_2 L_f L_g}, \frac{j}{L_f}\right\} \omega_{m_j} > \omega_{m_j}.$$

This yields a contradiction and so we must have

$$\limsup_{m\uparrow\infty} \frac{l_m - f(\boldsymbol{x}_*)}{\omega_m} < \infty$$

completing the proof. □

The next section describes an implementable version of the algorithms in Section 5.4 by constructing an easy set containing $\mathcal{X}_m$ as defined in Algorithm 5.1 or in Algorithm 5.2.

5.6. An ellipsoid algorithm for quasiconvex programming

As Section 5.4 shows, it is important to have an easy localization set to help performing a global search on $\mathcal{X}_m$ when one cannot resort to local information to compute an element of the normal cone. In this section we propose an adaptation of the ellipsoid algorithm which keeps track of an easy set which is guaranteed to include an optimal solution. This set is a *box*, or *hyperrectangle*, and its main properties are presented in the next subsection.

5.6.1. Ellipsoids and boxes

To start this section we first introduce some notation. For any compact set $\mathcal{C} \subseteq \mathbf{R}^s$ we denote by $\mathcal{R}(\mathcal{C})$ the smallest box containing $\mathcal{C}$, while $\mathcal{R}(\boldsymbol{l}, \boldsymbol{u}) := \{\boldsymbol{x} \in \mathbf{R}^s : \boldsymbol{l} \leq \boldsymbol{x} \leq \boldsymbol{u}\}$ is the box defined by $\boldsymbol{l}, \boldsymbol{u} \in \mathbf{R}^s$ with $\boldsymbol{l} \leq \boldsymbol{u}$. The following result characterizes $\mathcal{R}(\mathcal{C})$ for any compact set $\mathcal{C} \subseteq \mathbf{R}^s$.

Lemma 5.6.1. *For any compact set $\mathcal{C} \subseteq \mathbf{R}^s$ it follows that $\mathcal{R}(\mathcal{C}) = \mathcal{R}(\boldsymbol{l}, \boldsymbol{u})$ with $l_i := \min\{\boldsymbol{e}_i^\top \boldsymbol{x} : \boldsymbol{x} \in \mathcal{C}\}$ and $u_i := \max\{\boldsymbol{e}_i^\top \boldsymbol{x} : \boldsymbol{x} \in \mathcal{C}\}$ for each $i = 1, \ldots, s$ and $\boldsymbol{e}_i$ denoting the ith unit vector of $\mathbf{R}^s$.*

Proof: This is a trivial result and hence its proof is omitted. □

An illustration of Lemma 5.6.1 is given by Figure 5.9.

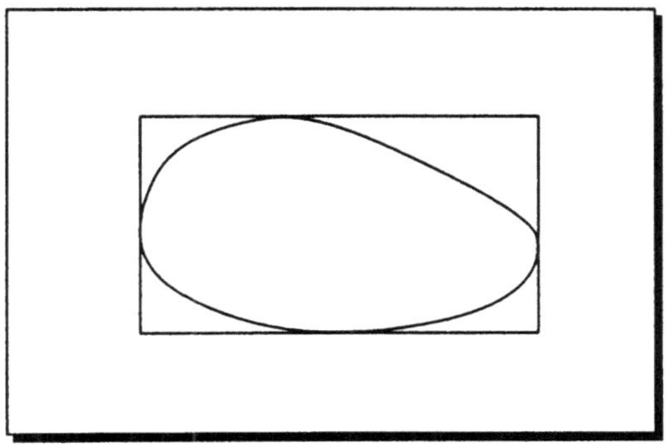

Figure 5.9: The smallest box including a given compact set

An immediate application of Lemma 5.6.1 is given by the following result.

Lemma 5.6.2. *$\mathcal{R}(\mathcal{E}(\boldsymbol{A}; \boldsymbol{a})) = \mathcal{R}(\boldsymbol{l}, \boldsymbol{u})$ with $l_i = a_i - \sqrt{a_{ii}}$ and $u_i = a_i + \sqrt{a_{ii}}$.*

Proof: Follows immediately from Lemma 5.6.1, (3.3) and (3.4). □

For the special case $\boldsymbol{A} = r^2 \boldsymbol{I}$ corresponding to the hypersphere usually taken as a starting ellipsoid for the ellipsoid method we obtain by Lemma 5.6.2 that $l_i = a_i - r$ and $u_i = a_i + r$.

Let $\mathcal{E}(\boldsymbol{A}; \boldsymbol{a})$ be an ellipsoid and $\mathcal{L}_h^{=}(\beta)$ with $h(\boldsymbol{x}) := \boldsymbol{a}^{\star\top}(\boldsymbol{x} - \boldsymbol{a})$ a cut of depth α with $-1 < \alpha < 1$. Recall from Chapter 3 that $\alpha := -\beta / \sqrt{\boldsymbol{a}^{\star\top} \boldsymbol{A} \boldsymbol{a}^\star}$. Observe that if $\alpha \leq -1$ then $\mathcal{E}(\boldsymbol{A}; \boldsymbol{a}) \cap \mathcal{L}_h^{\leq}(\beta) = \mathcal{E}(\boldsymbol{A}; \boldsymbol{a})$ and if $\alpha = 1$ the same intersection reduces to one point, $\boldsymbol{a} - \boldsymbol{A}\boldsymbol{a}^\star / \sqrt{\boldsymbol{a}^{\star\top} \boldsymbol{A} \boldsymbol{a}^\star}$. Moreover, if $\alpha > 1$ the intersection is empty. Before stating the next lemma, we introduce $\Lambda(\boldsymbol{x}, \boldsymbol{y}) := (\boldsymbol{x}^\top \boldsymbol{A} \boldsymbol{x})(\boldsymbol{y}^\top \boldsymbol{A} \boldsymbol{y}) - (\boldsymbol{x}^\top \boldsymbol{A} \boldsymbol{y})^2$. By Schwartz inequality it follows for any positive definite matrix $\boldsymbol{A}$ that $\Lambda(\boldsymbol{x}, \boldsymbol{y}) \geq 0$ for every pair of real vectors $\boldsymbol{x}$ and $\boldsymbol{y}$.

Lemma 5.6.3. *For any given vectors* $\boldsymbol{c}$ *and* $\boldsymbol{a}^\star$ *and scalar* β *such that* $-1 < \alpha < 1$ *it follows that*

$$\min\{\boldsymbol{c}^\top \boldsymbol{x} : \boldsymbol{x} \in \mathcal{E}(\boldsymbol{A};\boldsymbol{a}),\ \boldsymbol{a}^{\star\top}(\boldsymbol{x}-\boldsymbol{a}) \leq \beta\} = \begin{cases} \boldsymbol{c}^\top\boldsymbol{a} - \sqrt{\boldsymbol{c}^\top\boldsymbol{A}\boldsymbol{c}} & \text{if } \frac{\boldsymbol{c}^\top\boldsymbol{A}\boldsymbol{a}^\star}{\sqrt{\boldsymbol{c}^\top\boldsymbol{A}\boldsymbol{c}}\sqrt{\boldsymbol{a}^{\star\top}\boldsymbol{A}\boldsymbol{a}^\star}} \leq \alpha \\ \boldsymbol{c}^\top\boldsymbol{a} - \frac{\sqrt{1-\alpha^2}\sqrt{\Lambda(\boldsymbol{c},\boldsymbol{a}^\star)}+\alpha(\boldsymbol{c}^\top\boldsymbol{A}\boldsymbol{a}^\star)}{\sqrt{\boldsymbol{a}^{\star\top}\boldsymbol{A}\boldsymbol{a}^\star}} & \text{otherwise} \end{cases}.$$

Proof: We assume $\boldsymbol{c} \neq \boldsymbol{0}$ and $\boldsymbol{a}^\star \neq \boldsymbol{0}$ since any other case is not interesting.

$$(\Pi)\quad \min\{\boldsymbol{c}^\top \boldsymbol{x} : \boldsymbol{x} \in \mathcal{E}(\boldsymbol{A};\boldsymbol{a}),\ \boldsymbol{a}^{\star\top}(\boldsymbol{x}-\boldsymbol{a}) \leq \beta\}.$$

Since the above optimization problem (Π) is a nice convex program (Ref. [89]) we may resort to strong Lagrangian duality and address the problem

$$(\Delta)\quad \max\{\pi(\lambda) : \lambda \geq 0\}$$

with

$$\pi(\lambda) := \min\{\boldsymbol{c}^\top\boldsymbol{x} - \lambda(\beta - \boldsymbol{a}^{\star\top}(\boldsymbol{x}-\boldsymbol{a})) : \boldsymbol{x} \in \mathcal{E}(\boldsymbol{A};\boldsymbol{a})\}.$$

Applying (3.3) we obtain the following analytical expression

$$\pi(\lambda) = \boldsymbol{c}^\top\boldsymbol{a} - \beta\lambda - \sqrt{\boldsymbol{a}^{\star\top}\boldsymbol{A}\boldsymbol{a}^\star\lambda^2 + 2\boldsymbol{a}^{\star\top}\boldsymbol{A}\boldsymbol{c}\lambda + \boldsymbol{c}^\top\boldsymbol{A}\boldsymbol{c}}$$

Observe that function π is well defined and if $\boldsymbol{c}$ and $\boldsymbol{a}^\star$ are not colinear it is also continuously differentiable for every λ due to the positive definiteness of $\boldsymbol{A}$, since the radical is $(\boldsymbol{c}+\lambda\boldsymbol{a}^\star)^\top\boldsymbol{A}(\boldsymbol{c}+\lambda\boldsymbol{a}^\star) > 0$. Therefore we may compute the first derivative

$$\pi'(\lambda) = \frac{-\beta\ \sqrt{\boldsymbol{a}^{\star\top}\boldsymbol{A}\boldsymbol{a}^\star\lambda^2 + 2\boldsymbol{a}^{\star\top}\boldsymbol{A}\boldsymbol{c}\lambda + \boldsymbol{c}^\top\boldsymbol{A}\boldsymbol{c}} - \boldsymbol{a}^{\star\top}\boldsymbol{A}\boldsymbol{a}^\star\lambda - \boldsymbol{a}^{\star\top}\boldsymbol{A}\boldsymbol{c}}{\sqrt{\boldsymbol{a}^{\star\top}\boldsymbol{A}\boldsymbol{a}^\star\lambda^2 + 2\boldsymbol{a}^{\star\top}\boldsymbol{A}\boldsymbol{c}\lambda + \boldsymbol{c}^\top\boldsymbol{A}\boldsymbol{c}}}.$$

We also note that by computing the second derivative

$$\pi''(\lambda) = \frac{-\Lambda(\boldsymbol{a}^\star,\boldsymbol{c})}{\left(\sqrt{\boldsymbol{a}^{\star\top}\boldsymbol{A}\boldsymbol{a}^\star\lambda^2 + 2\boldsymbol{a}^{\star\top}\boldsymbol{A}\boldsymbol{c}\lambda + \boldsymbol{c}^\top\boldsymbol{A}\boldsymbol{c}}\right)^3}$$

and observing that it is negative we get a direct proof of the concavity of π. By solving $\pi'(\lambda) = 0$ we obtain

$$\lambda^\star = \frac{1}{\boldsymbol{a}^{\star\mathsf{T}} \boldsymbol{A} \boldsymbol{a}^\star} \left(\frac{\alpha}{\sqrt{1-\alpha^2}} \sqrt{\Lambda(\boldsymbol{a}^\star, \boldsymbol{c})} - \boldsymbol{a}^{\star\mathsf{T}} \boldsymbol{A} \boldsymbol{c} \right)$$

and

$$\pi(\lambda^\star) = \boldsymbol{c}^\mathsf{T} \boldsymbol{a} - \frac{\sqrt{\Lambda(\boldsymbol{a}^\star, \boldsymbol{c})} + \alpha \boldsymbol{a}^{\star\mathsf{T}} \boldsymbol{A} \boldsymbol{c}}{\sqrt{\boldsymbol{a}^{\star\mathsf{T}} \boldsymbol{A} \boldsymbol{a}^\star}}.$$

Now observe that the above value is the solution of $\max\{\pi(\lambda) : \lambda \in \mathbf{R}\}$, that is, the dual of $\min\{\boldsymbol{c}^\mathsf{T}\boldsymbol{x} : \boldsymbol{x} \in \mathcal{E}(\boldsymbol{A}; \boldsymbol{a}),\ \boldsymbol{a}^{\star\mathsf{T}}(\boldsymbol{x} - \boldsymbol{a}) = \beta\}$. So, if $\lambda^\star \geq 0$ it solves (Δ) and therefore (Π), otherwise 0 solves (Δ). We show now that the case $\boldsymbol{c}$ colinear with $\boldsymbol{a}^\star$ is covered by the same formulas. We note first that if $\boldsymbol{c} = t\boldsymbol{a}^\star$ for any $t \in \mathbf{R}$ then $\Lambda(\boldsymbol{a}^\star, \boldsymbol{c}) = 0$. Geometrically the solution is obvious. In fact, if $t > 0$ minimizing $\boldsymbol{c}^\mathsf{T}\boldsymbol{x}$ is, apart from t, equivalent to minimizing $\boldsymbol{a}^{\star\mathsf{T}}\boldsymbol{x}$ and therefore the active constraint is the ellipsoid and the result given by $\boldsymbol{c}^\mathsf{T}\boldsymbol{a} - \sqrt{\boldsymbol{c}^\mathsf{T}\boldsymbol{A}\boldsymbol{c}} = \pi(0)$. If, on the other hand, $t < 0$ minimizing $\boldsymbol{c}^\mathsf{T}\boldsymbol{x}$ is, apart from $-t$, equivalent to maximizing $\boldsymbol{a}^{\star\mathsf{T}}\boldsymbol{x}$ and therefore the active constraint is the hyperplane and the result is β. Suppose first $\boldsymbol{c} = t\boldsymbol{a}^\star$ for $t > 0$. Then $\boldsymbol{a}^{\star\mathsf{T}}\boldsymbol{A}\boldsymbol{c} = t\boldsymbol{a}^{\star\mathsf{T}}\boldsymbol{A}\boldsymbol{a}^\star > 0$ and the formula yields the right value, $\pi(0)$. If $\boldsymbol{c} = t\boldsymbol{a}^\star$ for $t < 0$ then $\boldsymbol{a}^{\star\mathsf{T}}\boldsymbol{A}\boldsymbol{c} = t\boldsymbol{a}^{\star\mathsf{T}}\boldsymbol{A}\boldsymbol{a}^\star < 0$. In this case the formula gives again the right value, namely $\pi(\lambda^\star)$ which one may verify to equal β. □

Concerning the special case of boxes, the following result can be easily derived from the previous result.

Lemma 5.6.4. *Let $\mathcal{E}(\boldsymbol{A}; \boldsymbol{a})$ be an ellipsoid and $\mathcal{L}_h^{=}(\beta)$ with $h(\boldsymbol{x}) := \boldsymbol{a}^{\star\mathsf{T}}(\boldsymbol{x} - \boldsymbol{a})$ a valid cut of depth α with $-1 < \alpha < 1$. Then $\mathcal{R}(\boldsymbol{l}, \boldsymbol{u}) := \mathcal{R}\left(\mathcal{E}(\boldsymbol{A}; \boldsymbol{a}) \cap \mathcal{L}_h^{\leq}(\beta)\right)$ is given by*

$$l_i := \begin{cases} a_i - \sqrt{a_{ii}} & \text{if } \boldsymbol{a}^{\star\mathsf{T}}\boldsymbol{a}_i \geq \alpha(\boldsymbol{a}^{\star\mathsf{T}}\boldsymbol{A}\boldsymbol{a}^\star)a_{ii} \\ a_i - \frac{\sqrt{1-\alpha^2}\sqrt{\boldsymbol{a}^{\star\mathsf{T}}\boldsymbol{A}\boldsymbol{a}^\star a_{ii} - (\boldsymbol{a}^{\star\mathsf{T}}\boldsymbol{a}_i)^2} + \alpha(\boldsymbol{a}^{\star\mathsf{T}}\boldsymbol{a}_i)}{\sqrt{\boldsymbol{a}^{\star\mathsf{T}}\boldsymbol{A}\boldsymbol{a}^\star}} & \text{otherwise} \end{cases}$$

and

$$u_i := \begin{cases} a_i + \sqrt{a_{ii}} & \text{if } \boldsymbol{a}^{\star\mathsf{T}}\boldsymbol{a}_i \leq -\alpha(\boldsymbol{a}^{\star\mathsf{T}}\boldsymbol{A}\boldsymbol{a}^\star)a_{ii} \\ a_i + \frac{\sqrt{1-\alpha^2}\sqrt{\boldsymbol{a}^{\star\mathsf{T}}\boldsymbol{A}\boldsymbol{a}^\star a_{ii} - (\boldsymbol{a}^{\star\mathsf{T}}\boldsymbol{a}_i)^2} - \alpha(\boldsymbol{a}^{\star\mathsf{T}}\boldsymbol{a}_i)}{\sqrt{\boldsymbol{a}^{\star\mathsf{T}}\boldsymbol{A}\boldsymbol{a}^\star}} & \text{otherwise} \end{cases}$$

with $\boldsymbol{a}_i$ the ith column of $\boldsymbol{A}$ for each $i = 1, \dots, s$.

Proof: Follows from Lemma 5.6.3 after replacing $\boldsymbol{c}$ by $\boldsymbol{e}_i$ for each l_i and by $-\boldsymbol{e}_i$ for each u_i. Remember in this last case to multiply the result by -1. □

For the proof of the next result result we use direct arguments.

Lemma 5.6.5. *Suppose $\mathcal{L}_{\bar{h}}^{\leq}(\beta)$ with $h(\boldsymbol{x}) := \boldsymbol{a}^{\star^{\mathsf{T}}}(\boldsymbol{x} - \boldsymbol{a})$ and $h(\boldsymbol{x}) \cap \mathcal{R}(\boldsymbol{l}, \boldsymbol{u}) \neq \emptyset$. Then $\mathcal{R}(\boldsymbol{l}', \boldsymbol{u}') := \mathcal{R}\left(\mathcal{L}_{\bar{h}}^{\leq}(\beta) \cap \mathcal{R}(\boldsymbol{l}, \boldsymbol{u})\right)$ is given by*

$$l_i' := \begin{cases} l_i & \text{if } a_i^\star \geq 0 \\ \max\left\{u_i + \frac{\beta'}{a_i^\star}, l_i\right\} & \text{otherwise} \end{cases}$$

and

$$u_i' := \begin{cases} u_i & \text{if } a_i^\star \leq 0 \\ \min\left\{l_i + \frac{\beta'}{a_i^\star}, u_i\right\} & \text{otherwise} \end{cases}$$

with $\beta' := \beta + \boldsymbol{a}^{\star^{\mathsf{T}}}\boldsymbol{a} - \sum_{j: a_j^\star < 0} a_j^\star u_j - \sum_{j: a_j^\star \geq 0} a_j^\star l_j$.

Proof: We give the following constructive proof of the result, which is based on the observation that if $\boldsymbol{l} = \boldsymbol{0}$ and $\boldsymbol{a}^{\star^{\mathsf{T}}}(\boldsymbol{l} - \boldsymbol{a}) \leq \beta$, i.e. $\beta' := \beta + \boldsymbol{a}^{\star^{\mathsf{T}}}\boldsymbol{a} \geq 0$, then the computation of the rectangle reduces to updating $\boldsymbol{u}$ as follows

$$u_i := \begin{cases} u_i & \text{if } a_i^\star \leq 0 \\ \min\left\{\frac{\beta'}{a_i^\star}, u_i\right\} & \text{otherwise} \end{cases}.$$

This amounts to comparing the present upper bounds with the intersections of the hyperplane and each coordinated axis, if this intersection occurs in the positive half of the corresponding axis.

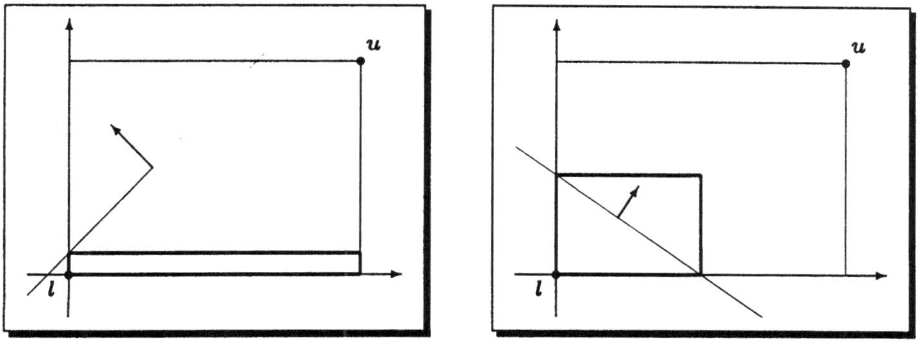

Figure 5.10: Updating an easy box

We now reduce the original problem to the above ideal situation. First we make sure that $\boldsymbol{a}^{\star\top}(\boldsymbol{l}' - \boldsymbol{a}) \leq \beta$ by performing the following variable exchange

$$x_i' := \begin{cases} x_i & \text{if } a_i^\star \geq 0 \\ -x_i & \text{otherwise} \end{cases} .$$

Observe this implies

$$l_i' := \begin{cases} l_i & \text{if } a_i^\star \geq 0 \\ -u_i & \text{otherwise} \end{cases} \text{ and } u_i' := \begin{cases} u_i & \text{if } a_i^\star \geq 0 \\ -l_i & \text{otherwise} \end{cases} .$$

In order to make $\boldsymbol{l}'' = \mathbf{0}$ another variable exchange is necessary, namely

$$x_i'' := \begin{cases} x_i' - l_i' & = & x_i - l_i & \text{if } a_i^\star \geq 0 \\ x_i' - l_i' & = & -x_i + u_i & \text{otherwise} \end{cases} .$$

The result follows now by updating this last box as listed above and reversing the pair of variable transformations. □

Figure 5.11 shows the smallest box that includes half of an ellipsoid, given by Lemma 5.6.4, and the smallest box including the intersection of a box and a halfspace, given by Lemma 5.6.5

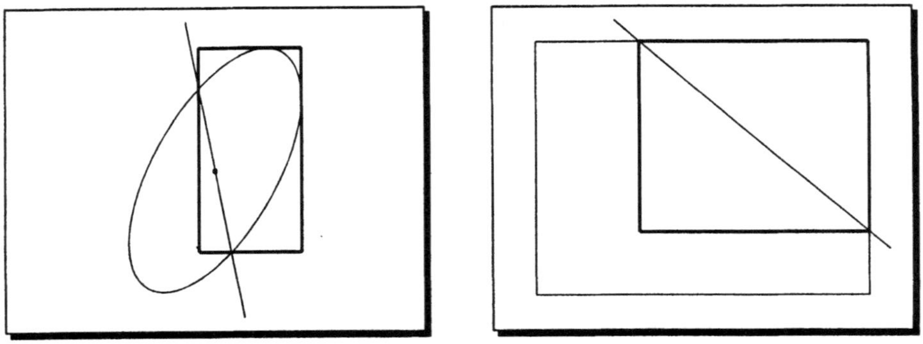

Figure 5.11: A pair of enclosing boxes

The following pair of simple results concludes this section. Observe that these results are needed in the next subsection.

Lemma 5.6.6. *$\mathcal{R}(\boldsymbol{l},\boldsymbol{u}) := \mathcal{R}(\boldsymbol{l}',\boldsymbol{u}') \cap \mathcal{R}(\boldsymbol{l}'',\boldsymbol{u}'')$ is given by $l_i := \max\{l_i', l_i''\}$ and $u_i := \min\{u_i', u_i''\}$.*

Proof: By writing $\mathcal{R}(\boldsymbol{l},\boldsymbol{u}) = \{\boldsymbol{x} \in \mathbf{R}^s : l_i' \le x_i \le u_i',\ l_i'' \le x_i \le u_i'',\ 1 \le i \le s\}$ the result becomes obvious. □

Lemma 5.6.7. *$\mathcal{R}(\mathcal{A}\cap\mathcal{B}) \subseteq \mathcal{R}(\mathcal{A})\cap\mathcal{R}(\mathcal{B}) \subseteq \mathcal{R}(\mathcal{A}\cup\mathcal{B})$ holds for each pair of compact sets $\mathcal{A}$ and $\mathcal{B}$.*

Proof: Observe that $\mathcal{X} \subseteq \mathcal{Y}$ implies that $\mathcal{R}(\mathcal{X}) \subseteq \mathcal{R}(\mathcal{Y})$. The result follows now from $\mathcal{A}\cap\mathcal{B} \subseteq \mathcal{A}$, $\mathcal{A}\cap\mathcal{B} \subseteq \mathcal{B}$, $\mathcal{A} \subseteq \mathcal{A}\cup\mathcal{B}$ and $\mathcal{B} \subseteq \mathcal{A}\cup\mathcal{B}$. □

5.6.2. Constructing a localization box

Our goal is to construct a box that in each iteration m of the ellipsoid algorithm, as described in Chapter 3, includes $\bigcap_{k=0}^{m-1}\left(\mathcal{E}(\boldsymbol{A}_k;\boldsymbol{a}_k) \cap \mathcal{L}_{\bar{h}_k}^{\le}(\beta_k)\right)$ and is as small as possible. Using the tools developed in the previous section we may modify the ellipsoid algorithm as follows.

Step 0 let $m := 0$, $\boldsymbol{A}_m := r^2\boldsymbol{I}$ and $\mathcal{R}(\boldsymbol{l}_m, \boldsymbol{u}_m) :=$ $\boxed{\mathcal{R}(\mathcal{E}(\boldsymbol{A}_m; \boldsymbol{a}_m))}$;

Step 1 *if* $\boldsymbol{a}_m$ satisfies some stopping criteria

then stop

else proceed to **Step 2**;

Step 2 *if* $n(\boldsymbol{a}_m) > r$ *then* $\boxed{\text{apply a norm cut}}$

else if $g(\boldsymbol{a}_m) > 0$ *then* $\boxed{\text{apply a constraint cut}}$

else $\boxed{\text{apply an objective cut}}$;

Step 3 $\mathcal{R}(\boldsymbol{l}_{m+1}, \boldsymbol{u}_{m+1}) :=$

$$\boxed{\mathcal{R}\left(\mathcal{E}(\boldsymbol{A}_m; \boldsymbol{a}_m) \cap \mathcal{L}^{\leq}_{\bar{h}_m}(\beta_m)\right) \cap \mathcal{R}\left(\mathcal{R}(\boldsymbol{l}_m, \boldsymbol{u}_m) \cap \mathcal{L}^{\leq}_{\bar{h}_m}(\beta_m)\right)};$$

$\boxed{\text{update the ellipsoid}}$;

let $m := m + 1$ and return to **Step 1**

Algorithm 5.3: The modified ellipsoid algorithm for quasiconvex programming

For the definition of the function n we refer to Chapter 3.

The above algorithm requires the specification of six procedures, marked as framed statements. The application of a norm cut is carried out by Algorithm 3.12, while the application of a constraint cut follows Algorithm 3.8. The update of the ellipsoid resorts to Algorithm 3.9 and to Algorithm 3.10 in order to implement the formulas (3.6) and (3.7) in a numerically robust way. The application of an objective cut follows Algorithm 5.1 or Algorithm 5.2.

Finally, two procedures return boxes. The first corresponds to the computation of the smallest box containing the initial ellipsoid and this can be implemented by the remark after Lemma 5.6.2 as in Algorithm 5.4. Observe that the index "m", which equals zero at this stage, is dropped.

for $j = 1$ *to* s *do* $l_j := a_j - r$ and $u_j := a_j + r$

Algorithm 5.4: Constructing the first box

The second corresponds to updating the current box and is listed in Algorithm 5.5. Observe that θ is used in Algorithm 5.5 as an "auxiliary variable" and that this procedure modifies the current box. Moreover, **Step 1** of this algorithm corresponds to applying Lemma 5.6.5 while **Step 2** simultaneously applies Lemma 5.6.4 and Lemma 5.6.6.

Step 1 $\theta := \beta + \boldsymbol{a}^{\star\top}\boldsymbol{a}$;
for $j := 1$ *to* s *do*
if $a_j^\star < 0$ *then* $\theta := \theta - a_j^\star u_j$
else $\theta := \theta - a_j^\star l_j$;
for $i := 1$ *to* s *do*
if $a_i^\star < 0$ *then if* $u_i + \theta/a_i^\star > l_i$ *then* $l_i := u_i + \theta/a_i^\star$;
if $a_i^\star > 0$ *then if* $l_i + \theta/a_i^\star < u_i$ *then* $u_i := l_i + \theta/a_i^\star$;
Step 2 *for* $i := 1$ *to* s *do*
if $\boldsymbol{a}^{\star\top}\boldsymbol{a}_i \geq \alpha(\boldsymbol{a}^{\star\top}\boldsymbol{A}\boldsymbol{a}^\star)a_{ii}$
then $\theta := a_i - \sqrt{a_{ii}}$
else $\theta := a_i - \frac{\sqrt{1-\alpha^2}\sqrt{\boldsymbol{a}^{\star\top}\boldsymbol{A}\boldsymbol{a}^\star a_{ii} - (\boldsymbol{a}^{\star\top}\boldsymbol{a}_i)^2} + \alpha(\boldsymbol{a}^{\star\top}\boldsymbol{a}_i)}{\sqrt{\boldsymbol{a}^{\star\top}\boldsymbol{A}\boldsymbol{a}^\star}}$;
if $\theta > l_i$ *then* $l_i := \theta$;
if $\boldsymbol{a}^{\star\top}\boldsymbol{a}_i \leq -\alpha(\boldsymbol{a}^{\star\top}\boldsymbol{A}\boldsymbol{a}^\star)a_{ii}$
then $\theta := a_i + \sqrt{a_{ii}}$
else $\theta := a_i + \frac{\sqrt{1-\alpha^2}\sqrt{\boldsymbol{a}^{\star\top}\boldsymbol{A}\boldsymbol{a}^\star a_{ii} - (\boldsymbol{a}^{\star\top}\boldsymbol{a}_i)^2} - \alpha(\boldsymbol{a}^{\star\top}\boldsymbol{a}_i)}{\sqrt{\boldsymbol{a}^{\star\top}\boldsymbol{A}\boldsymbol{a}^\star}}$;
if $\theta < u_i$ *then* $u_i := \theta$

Algorithm 5.5: Updating the box

The following lemma establishes $\mathcal{R}(\boldsymbol{l}_m, \boldsymbol{u}_m)$ as a suitable set to be searched for an element of $\mathcal{X}_m$.

Lemma 5.6.8. *The box $\mathcal{R}(\boldsymbol{l}_m, \boldsymbol{u}_m)$ constructed by the modified ellipsoid algorithm includes $\bigcap_{k=0}^{m-1}\left(\mathcal{E}(\boldsymbol{A}_k; \boldsymbol{a}_k) \cap \mathcal{L}_{\bar{h}_k}^{\leq}(\beta_k)\right)$ at each iteration m.*

Proof: We proof this result by induction. For $m = 1$ the result follows from Lemma 5.6.7 since $\mathcal{R}(\boldsymbol{l}_1, \boldsymbol{u}_1) := \mathcal{R}\left(\mathcal{E}(\boldsymbol{A}_0; \boldsymbol{a}_0) \cap \mathcal{L}_{\bar{h}_0}^{\leq}(\beta_0)\right) \cap \mathcal{R}\left(\mathcal{R}(\boldsymbol{l}_0, \boldsymbol{u}_0) \cap \mathcal{L}_{\bar{h}_0}^{\leq}(\beta_0)\right)$

and $\mathcal{R}(\boldsymbol{l}_0, \boldsymbol{u}_0) := \mathcal{R}(\mathcal{E}(\boldsymbol{A}_0; \boldsymbol{a}_0))$. Suppose that it holds for m, i.e. $\mathcal{R}(\boldsymbol{l}_{m+1}, \boldsymbol{u}_{m+1}) := \mathcal{R}\left(\mathcal{E}(\boldsymbol{A}_m; \boldsymbol{a}_m) \cap \mathcal{L}^{\leq}_{\bar{h}_m}(\beta_m)\right) \cap \mathcal{R}\left(\mathcal{R}(\boldsymbol{l}_m, \boldsymbol{u}_m) \cap \mathcal{L}^{\leq}_{\bar{h}_m}(\beta_m)\right)$. Since by the induction hypothesis it follows that $\mathcal{R}(\boldsymbol{l}_m, \boldsymbol{u}_m) \supseteq \bigcap_{k=0}^{m-1} (\mathcal{E}(\boldsymbol{A}_k; \boldsymbol{a}_k) \cap \underline{\mathcal{H}}(\boldsymbol{d}_k, \beta_k))$ then the result is proven by applying Lemma 5.6.7. □

We finish this subsection by pointing out that by Lemma 5.6.8 and Theorem 5.4.1 the set $\mathcal{R}(\boldsymbol{l}_m, \boldsymbol{u}_m)$ includes $\mathcal{X}_m$ and it is suitable to be searched. A possible search strategy is the following. Every time the search procedure generates a point $\boldsymbol{y}$ in $\mathcal{R}(\boldsymbol{l}_m, \boldsymbol{u}_m)$ this point is tested for membership of $\mathcal{X}_m$. If such a point $\boldsymbol{y} \in \mathcal{X}_m$ is found then Lemma 5.4.1 reduces the search space to the segment $]\boldsymbol{y}, \boldsymbol{a}_m[$.

Suppose now that such a search method is implemented by means of grid search and that a grid is laid over $\mathcal{R}(\boldsymbol{l}_m, \boldsymbol{u}_m)$ with coordinate steps of $\delta > 0$. If such a grid finds no point in $\mathcal{X}_m$ then it follows that either $\mathcal{X}_m$ is empty and hence $\boldsymbol{a}_m$ is optimal, or an error is incurred. We now try to estimate this error. If $\boldsymbol{y} \in \mathcal{X}_m$ exists, it can not be at a bigger distance from an examined point in the grid, say $\boldsymbol{w}$ with $f(\boldsymbol{w}) \geq \ell_m$, than $\|\boldsymbol{w} - \boldsymbol{y}\|_\infty$. By the hypothesis on the grid size it follows that $\|\boldsymbol{w} - \boldsymbol{y}\|_\infty \leq \delta$ and hence if we also assume that f is Lipschitz continuous with Lipschitz constant $L_f > 0$ we obtain the following inequality

$$|f(\boldsymbol{w}) - f(\boldsymbol{y})| \leq L_f \|\boldsymbol{w} - \boldsymbol{y}\|_2 \leq L_f \sqrt{s} \|\boldsymbol{w} - \boldsymbol{y}\|_\infty \leq L_f \delta \sqrt{s}.$$

Hence we conclude that the difference between the objective function value of any point in the grid and the value of any possible missed improving point can not exceed $L_f \delta \sqrt{s}$. In particular $\ell_m - f(\boldsymbol{x}_*) \leq L_f \delta / \sqrt{s}$ and hence if a Lipschitz constant is known the value δ can be chosen accordingly.

5.6.3. New cuts

So far the box has just been updated in each iteration of the ellipsoid algorithm. We use it now in order to modify the progress of the algorithm itself. This is achieved by introducing two new classes of cuts, both originated by the box.

Each new cut is performed only if the volume of the resulting ellipsoid becomes smaller than the volume of the ellipsoid that would otherwise be created by the algorithm.

We now recall from Chapter 3 that

$$\mu_m := \frac{\operatorname{vol}(\mathcal{E}(\boldsymbol{A}_{m+1}; \boldsymbol{a}_{m+1}))}{\operatorname{vol}(\mathcal{E}(\boldsymbol{A}_m; \boldsymbol{a}m))} = \sqrt{\delta_m^s(1-\sigma_m)}$$

with $\boldsymbol{A}_{m+1} := \delta_m(\boldsymbol{A}_m - \sigma_m \boldsymbol{b}_m \boldsymbol{b}_m^{\mathsf{T}})$ and $\boldsymbol{b}_m := \boldsymbol{A}_m \boldsymbol{a}_m^\star / \sqrt{\boldsymbol{a}_m^{\star\mathsf{T}} \boldsymbol{A}_m \boldsymbol{a}_m^\star}$. Moreover, $\boldsymbol{a}_m^\star$ is the normal to the cut and δ_m and σ_m are updating values given by the type of cut used. Thus, the new cut is applied instead of the old cut if the corresponding μ_m is the smaller of the two. Consequently, since the convergence proof given in Chapter 3 is based on the decreasing nature of $\operatorname{vol}(\mathcal{E}(\boldsymbol{A}_m; \boldsymbol{a}_m))$ and this is proportional to $\det(\boldsymbol{A}_m)$ the same proof applies when solving convex programs by means of the ellipsoid algorithm with the new cuts. In fact, the new algorithm generates a sequence $\operatorname{vol}(\mathcal{E}(\boldsymbol{A}_m; \boldsymbol{a}_m))$ which is upper bounded by the sequence of volumes created by the original algorithm.

In the remainder we drop the iterate index m for the sake of notational convenience.

5.6.4. Box cuts

A *box cut* is just a deep cut which is applied if the center $\boldsymbol{a}$ of the current ellipsoid is outside the current box.

Let us define

$$\Delta_i := \begin{cases} a_i - u_i & \text{if } a_i > u_i \\ l_i - a_i & \text{if } a_i < l_i \\ 0 & \text{otherwise} \end{cases}$$

and assume that $\max_{1 \le i \le s} \Delta_i > 0$, i.e. the center of the ellipsoid is outside the box. In this case let

$$\boldsymbol{a}_i^\star := \begin{cases} \boldsymbol{e}_i & \text{if } a_i > u_i \\ -\boldsymbol{e}_i & \text{if } a_i < l_i \end{cases} \quad \text{and } \beta_i := \begin{cases} u_i - a_i & \text{if } a_i > u_i \\ a_i - l_i & \text{if } a_i < l_i \end{cases}$$

for each $1 \leq i \leq s$ such that $\Delta_i > 0$. Recall that $\alpha_i := -\beta_i/\sqrt{\boldsymbol{a}_i^{\star\mathsf{T}} \boldsymbol{A} \boldsymbol{a}_i^\star} = \Delta_i/\sqrt{a_{ii}}$.

Each of these cuts is deep due to $\Delta_i > 0$ and valid due to the fact that optimal solutions are contained in the current box.

The candidate box cut is now given by ℓ if $\Delta_\ell/\sqrt{a_{\ell\ell}} = \max_{i:\Delta_i>0} \Delta_i/\sqrt{a_{ii}}$.

5.6.5. Parallel cuts

A parallel cut is defined by a pair of hyperplanes with the same normal. In this case we like to compute the smallest ellipsoid containing $\mathcal{L}_{\bar{h}}^{\geq}(\beta_1) \cap \mathcal{E}(\boldsymbol{A};\boldsymbol{a}) \cap \mathcal{L}_{\bar{h}}^{\leq}(\beta_2)$ with $\beta_1 \leq \boldsymbol{a}^{\star\mathsf{T}}\boldsymbol{x} \leq \beta_2$ the slice (pair of parallel cuts) of interest.

From Ref. [14] we know that an ellipsoid can be computed containing $\mathcal{L}_{\bar{h}}^{\geq}(\beta_1) \cap \mathcal{E}(\boldsymbol{A};\boldsymbol{a}) \cap \mathcal{L}_{\bar{h}}^{\leq}(\beta_2)$ and that ellipsoid has smaller volume than the current one if and only if $\alpha_1\alpha_2 < 1/s$ and $\alpha_1 \leq -\alpha_2 \leq 1$ with $\alpha_1 := \beta_1/\sqrt{\boldsymbol{a}^{\star\mathsf{T}}\boldsymbol{A}\boldsymbol{a}^\star}$ and $\alpha_2 := -\beta_2/\sqrt{\boldsymbol{a}^{\star\mathsf{T}}\boldsymbol{A}\boldsymbol{a}^\star}$. Moreover, this ellipsoid is given by the usual updating formulas, Ref. [14, 48], if the updating values are given as in Ref. [14] by

$$
\begin{aligned}
\delta &:= \frac{s^2}{s^2-1}\left(1 - \frac{\alpha_1^2+\alpha_2^2-\rho/s}{2}\right),\\
\sigma &:= \frac{2}{s+1}\left(\frac{s}{2} + \frac{1-\alpha_1\alpha_2-\rho/2}{(\alpha_1-\alpha_2)^2}\right) \quad \text{and}\\
\tau &:= \sigma\frac{\alpha_1-\alpha_2}{2}
\end{aligned}
$$

with $\rho = \sqrt{4(1-\alpha_1^2)(1-\alpha_2^2)+s^2(\alpha_1^2-\alpha_2^2)^2}$.

We implement parallel cuts as follows. For each $1 \leq i \leq s$ we take $\boldsymbol{a}^\star := \boldsymbol{e}_i$, $\beta_1 := l_i$ and $\beta_2 := u_i$. First, it is checked[4] if $\alpha_1\alpha_2 < 1/s$ and $\alpha_2 \leq -\alpha_1 \leq 1$. For each (if any) parallel cut that fulfills the previous test, $\mu = \sqrt{\delta^s(1-\sigma)}$ is computed and the candidate parallel cut is taken as the one that passes the previous test and minimizes μ among $1 \leq i \leq s$.

5.6.6. Modified algorithm

We now describe the modified algorithm in detail.

[4] Observe that $\alpha_1 = -l_i/\sqrt{a_{ii}}$ and $\alpha_2 = u_i/\sqrt{a_{ii}}$. Hence $\alpha_1\alpha_2 = -l_iu_i/a_{ii}$.

This algorithm requires the specification of seven procedures, marked as framed statements. The application of a norm cut is done by means of Algorithm 3.12, while the application of a constraint cut follows Algorithm 3.8. The update of the ellipsoid resorts to Algorithm 3.9 and to Algorithm 3.10 in order to implement the formulas (3.6) and (3.7) in a numerically robust way. The application of an objective cut follows Algorithm 5.1 or Algorithm 5.2. Observe that for this algorithm, the previously mentioned cuts are not yet effectively applied. In fact, a cut is only effectively applied by the procedure that updates the ellipsoid. The selection of the cut is carried out by Algorithm 5.7.

Step 0 let $m := 0$, $\boldsymbol{A}_0 := r^2\boldsymbol{I}$ and $\mathcal{R}(\boldsymbol{l}_0, \boldsymbol{u}_0) :=$ $\boxed{\mathcal{R}(\mathcal{E}(\boldsymbol{A}_0; \boldsymbol{a}_0))}$;

Step 1 *if* $\boldsymbol{a}_m$ satisfies some stopping criteria
then stop
else proceed to **Step 2**;

Step 2 *if* $n(\boldsymbol{a}_m) > r$ *then* $\boxed{\text{apply a norm cut}}$
else if $g(\boldsymbol{a}_m) > 0$ *then* $\boxed{\text{apply a constraint cut}}$
else $\boxed{\text{apply an objective cut}}$;

compute $\mu_m := \sqrt{\delta_m^s(1 - \sigma_m)}$;

Step 3 $\mathcal{R}(\boldsymbol{l}_{m+1}, \boldsymbol{u}_{m+1}) :=$

$$\boxed{\mathcal{R}\left(\mathcal{E}(\boldsymbol{A}_m; \boldsymbol{a}_m) \cap \mathcal{L}_{\bar{h}_m}^{\leq}(\beta_m)\right) \cap \mathcal{R}\left(\mathcal{R}(\boldsymbol{l}_m, \boldsymbol{u}_m) \cap \mathcal{L}_{\bar{h}_m}^{\leq}(\beta_m)\right)};$$

Step 4 $\boxed{\text{apply the best cut}}$;
$\boxed{\text{update the ellipsoid}}$;
let $m := m + 1$ and return to **Step 1**

Algorithm 5.6: The Ellipsoid & Box algorithm

The procedures dealing with boxes are given respectively by Algorithm 5.4 and Algorithm 5.5.

Observe that the cut selected in **Step 2** is always used to update the box before trying to use the updated box to generate a new cut and to update the ellipsoid. Moreover, due to the nature of the new cuts (box or parallel) none of them can make further changes in the current box. Also note that the number of iterations, i.e. the number of ellipsoid updates performed by the algorithm, equals the number of objective plus constraint plus norm cuts, and not the previous number plus box plus parallel cuts.

Notice that the computation of μ can be efficiently performed. In fact, let $\delta_c := \frac{s^2}{s^2-1}$ and $\sigma_c := \frac{2}{s+1}$ be the updating values of a central cut. We have

$$\mu_c := \sqrt{\delta_c^s(1-\sigma_c)} = s^s\sqrt{\frac{s-1}{(s+1)(s^2-1)^s}}.$$

Let now $\delta_d = \delta_c(1-\alpha^2)$ and $\sigma_d = \sigma_c\frac{1+s\alpha}{1+\alpha}$ be the same values in case of a deep cut of depth α. In this case we obtain

$$\mu_d := \sqrt{\delta_d^s(1-\sigma_d)} = \mu_c\sqrt{\frac{1-\alpha}{1+\alpha}(1-\alpha^2)^s}.$$

Finally, let us consider the case of parallel cuts with $\delta_p = \delta_c\left(1 - \frac{\alpha_1^2+\alpha_2^2-\rho/s}{2}\right)$ and $\sigma_p = \sigma_c\left(\frac{s}{2} + \frac{1-\alpha_1\alpha_2-\rho/2}{(\alpha_1-\alpha_2)^2}\right)$ the updating values. The values ρ, α_1 and α_2 are defined in Subsection 5.6.5. For parallel cuts we get

$$\mu_p := \sqrt{\delta_p^s(1-\sigma_p)} = \mu_c\sqrt{\left(\frac{\theta+\rho/s}{2}\right)^s \frac{\rho-\theta}{(s-1)(\alpha_1-\alpha_2)^2}}$$

with $\theta := 2-\alpha_1^2-\alpha_2^2$. Therefore, $\mu_d < \mu_c$ if and only if $\alpha > 0$, and $\mu_p < \mu_d$ if and only if

$$\left(\frac{\theta+\rho/s}{2}\right)^s \frac{\rho-\theta}{(s-1)(\alpha_1-\alpha_2)^2} < \frac{1-\alpha}{1+\alpha}(1-\alpha^2)^s.$$

This avoids computing μ_c and saves the computation of a square root. The application of the best cut is now detailed in Algorithm 5.7.

if [there exists a candidate box cut and this cut is ℓ]
  then if $\Delta_\ell/\sqrt{a_{\ell\ell}} > \alpha$
    then let $\alpha_m := \Delta_\ell/\sqrt{a_{\ell\ell}}$;
      if $a_\ell > u_\ell$ *then* let $\boldsymbol{a}_m^\star := \boldsymbol{e}_\ell$
        else let $\boldsymbol{a}_m^\star := -\boldsymbol{e}_\ell$;
      let $\delta_m := \frac{s^2(1-\alpha_m^2)}{s^2-1}$, $\sigma_m := \frac{2(1+s\alpha_m)}{(s+1)(1+\alpha_m)}$ and $\tau_m := \frac{1+s\alpha_m}{s+1}$;
if [there exists a candidate parallel cut]
  then if $\left(\frac{\theta_m+\rho_m/s}{2}\right)^s \frac{\rho_m-\theta_m}{(s-1)(\alpha_{1_m}-\alpha_{2_m})^2} < \frac{1-\alpha_m}{1+\alpha_m}(1-\alpha_m^2)^s$
    then let $\delta_m := \frac{s^2}{s^2-1}\left(1 - \frac{\alpha_{1_m}^2+\alpha_{2_m}^2-\rho_m/s}{2}\right)$;
      let $\sigma_m := \frac{2}{s+1}\left(\frac{s}{2} + \frac{1-\alpha_{1_m}\alpha_{2_m}-\rho_m/2}{(\alpha_{1_m}-\alpha_{2_m})^2}\right)$;
      let $\tau_m := \sigma\frac{\alpha_{1_m}-\alpha_{2_m}}{2}$

Algorithm 5.7: Application of the best available cut

The selection of the candidate box cut, if any, is explained in Subsection 5.6.4 while the selection of the candidate parallel cut, if any, is explained in Subsection 5.6.5.

Observe that none of the new cuts requires function evaluations and can be implemented with very little computational effort. In fact, it is easy to see that the total effort per iteration (despising possible need for global search) is of the same order as for Algorithm 3.11.

5.7. Improving the stopping criteria

Subsection 3.3 discusses criteria to ensure that a given precision has been met by Algorithm 3.11. The lower bounds used to derive the stopping rule presented in Subsection 3.3 are based on the subgradient inequality, which can be replaced by the lower subgradient inequality in case of a lower subdifferentiable function. According to this inequality

$$f(\boldsymbol{x}) \geq f(\boldsymbol{y}) + \boldsymbol{a}^{\star\mathsf{T}}(\boldsymbol{x}-\boldsymbol{y})$$

holds for every $\boldsymbol{x}$ in the domain of f (or in a lower level set in the case of a lower subdifferentiable function) with $\boldsymbol{a}^\star$ a (lower) subgradient of f at the point $\boldsymbol{y}$.

The lower bound derived in Subsection 3.3 is the value of

$$\min\{f(\boldsymbol{a}_m) + \boldsymbol{a}_m^{\star\mathsf{T}}(\boldsymbol{x} - \boldsymbol{a}_m) : \boldsymbol{x} \in \mathcal{E}(\boldsymbol{A}_m; \boldsymbol{a}_m)\}$$

which is given by (3.3).

An immediate way to strengthen this value is to compute also

$$\min\{f(\boldsymbol{a}_m) + \boldsymbol{a}_m^{\star\mathsf{T}}(\boldsymbol{x} - \boldsymbol{a}_m) : \boldsymbol{x} \in \mathcal{R}(\boldsymbol{l}_m, \boldsymbol{u}_m)\},$$

which can be obtained by inspecting the sign of each coordinate of $\boldsymbol{a}_m^\star$, and taking the maximum of this two values as the lower bound.

Moreover we observe that each of the s values

$$\min\{f(\boldsymbol{a}_m) + \boldsymbol{a}_m^{\star\mathsf{T}}(\boldsymbol{x} - \boldsymbol{a}_m) : \boldsymbol{x} \in \mathcal{E}(\boldsymbol{A}_m; \boldsymbol{a}_m) \cap \mathcal{H}^{\leq}(\boldsymbol{e}_i, u_i)\}$$

and each of the s values

$$\min\{f(\boldsymbol{a}_m) + \boldsymbol{a}_m^{\star\mathsf{T}}(\boldsymbol{x} - \boldsymbol{a}_m) : \boldsymbol{x} \in \mathcal{E}(\boldsymbol{A}_m; \boldsymbol{a}_m) \cap \mathcal{H}^{\geq}(\boldsymbol{e}_i, l_i)\}$$

is still a valid lower bound and so the maximum of this $2s$ values yields a lower bound as well.

VI Quasiconvexity in Location

This chapter complements Chapter 4 by using Chapter 5 in order to enlarge the class of location problems that can be solved by the ellipsoid method. Section 6.1 justifies the reason for considering quasiconvex disutility functions in location modeling and Section 6.2 details a quasiconvex location model. Finally, Section 6.3 presents some computational results.

6.1. Introduction

The location model considered in this chapter belongs to the class of problems described in Section 4.3. In fact, reading this chapter should be preceded by reading Section 4.1, Section 4.2 and Section 4.3.

We consider now as disutility functions associated with the demand points $\boldsymbol{d}_i$ (as introduced in Section 4.1) functions $\varphi_i : \mathbf{R}_+^n \longrightarrow \mathbf{R}_+$, $i = 1, \ldots, n$ which are nondecreasing and *quasiconvex.*

To cite just a few evidences of the usefulness of quasiconvex disutility functions in location modeling, we mention those instances dealing with transportation costs as functions of the traveled distance. In many applications, such costs offer "discounts per quantity", i.e. the longer the trip the cheaper the price per unit of length, say kilometer. Also economic applications often consider *concave* cost functions. An important subset of nonincreasing concave functions is composed of quasiconvex functions. A last example is provided by the Stochastic Queue Location Problems proposed by Zhang in Ref. [120]. These models have, under mild assumptions, quasiconvex disutility functions.

6.2. A quasiconvex location model

As mentioned in the previous section we assume in this chapter the disutility functions to be nondecreasing quasiconvex functions. Since by Chapter 5 the use of general quasiconvex functions may lead to non tractable models we impose some restrictions. We only consider functions like the ones given in Lemma 5.3.4. However, in order to enable some modeling freedom within this class we proceed as in Section 4.5 by letting each φ_i be given as the maximum of a finite number of known quasiconvex functions which are continuously differentiable on a open convex set $\mathcal{S} \supseteq \mathbf{R}^n_+$, say

$$\varphi_i(z) := \max_{1 \le j \le \kappa_i} \varphi_{i_j}(z).$$

Contrary to Ref. [40] the disutility functions considered in this section do not need to be differentiable. Nevertheless, as the remainder of this section shows the analysis of this model remains quite simple.

As in Section 4.5 we first provide a more appealing form to the model by exploiting the associative property of the finite maximum operator.

Let us take $\kappa := \sum_{i=1}^n \kappa_i$ and reorder the indices of φ_{i_j} in such a way that

$$\varphi(z) = \max_{1 \le i \le n} \max_{1 \le j \le \kappa_i} \varphi_{i_j}(z) = \max_{1 \le k \le \kappa} \varphi_k(z)$$

Our objective function is now given by $f(x) := \max_{1 \le k \le \kappa} f_k(x)$ with $f_k(x) := \varphi_k(\gamma(x))$.

The following theorem is now an immediate consequence of Lemma 5.3.4.

Theorem 6.2.1. *If $C(x) \subseteq \mathbf{R}^s$ denotes the set*

$$\operatorname{conv}\left(\bigcup_{k \in K(x)} \left(\sum_{j=1}^n \tfrac{\partial \varphi_k}{\partial z_j}(\gamma(x)) \partial \gamma_{\mathcal{G}_j}(x - d_j)\right)\right)$$

with $K(x) := \{k : \varphi_k(\gamma(x)) = \varphi(\gamma(x)),\ k = 1, \ldots, \kappa\}$ the so-called active set then x_0 belongs to Γ_f if and only if $\mathbf{0} \in C(x_0)$.

Observe the rather surprising similarity between Theorem 6.2.1 and Theorem 4.5.1. This implies that Example 4.5.1 and Example 4.5.2, discussed in Section 4.5, can also be used in this more general setting. Moreover, in the planar ($s = 2$) case the same algorithms as given in Section 4.6 can be applied to implement the above decision problem whether x_0 belongs to Γ_f or not.

As noticed in Chapter 5 the set Γ_f might contain nonoptimal points. In order to exhibit a case for which Γ_f coincides with the set of optimal points we introduce the following set of functions.

It is not difficult to verify for $\varphi_k : \mathbf{R}^n \longrightarrow \mathbf{R}$ nondecreasing on $\mathbf{R}^n_+$ and quasiconvex on the open convex set $\mathcal{S}$ with $\mathbf{R}^n_+ \subseteq \mathcal{S}$ that the function f_k is quasiconvex on $\mathbf{R}^s$. Since quasiconvexity is preserved under the max operator this yields that the objective function $f = \max_{1 \leq k \leq \kappa} f_k$ of (L) is also quasiconvex on $\mathbf{R}^s$. It is now possible to prove the following result.

Lemma 6.2.1. *If for every $1 \leq k \leq \kappa$ the function $\varphi_k : \mathbf{R}^n \longrightarrow \mathbf{R}$ is nondecreasing on $\mathbf{R}^n_+$ and differentiable and quasiconvex on an open convex set $\mathcal{S}$ with $\mathbf{R}^n_+ \subseteq \mathcal{S}$ and the gradient $\nabla\varphi_k(z)$ contains at least one positive component for every $z \in \mathcal{S}$ then the set Γ_f of stationary points coincides with the set of optimal solutions of (L).*

Proof: Clearly every optimal solution of (L) belongs to Γ_f. To prove the reverse we observe the following. For every $1 \leq k \leq \kappa$ it follows by the nonnegativity of $\nabla\varphi_k(z)$ for every $z \in \mathcal{S}$ and the convexity of $\gamma_{\mathcal{G}_j}$, $1 \leq j \leq n$, that

$$\begin{aligned} f_k'(x;y) &= \sum_{j=1}^{n} \tfrac{\partial\varphi_k}{\partial z_j}(\gamma(x))\gamma_{\mathcal{G}_j}'(x - d_j; y) \\ &\leq \sum_{j=1}^{n} \tfrac{\partial\varphi_k}{\partial z_j}(\gamma(x))(\gamma_{\mathcal{G}_j}(x + y - d_j) - \gamma_{\mathcal{G}_j}(x - d_j)) \\ &= \varphi_k'(\gamma(x); \gamma(x+y) - \gamma(x)) \end{aligned}$$

for every $y \in \mathbf{R}^s$. Hence for $x_0 \in \Gamma_f$ and $y \in \mathbf{R}^s$ fixed this implies by Lemma 2.5.3 that

$$\begin{aligned} 0 \leq f'(x_0; y) &= \max\{f_k'(x_0; y) : i \in I(x_0)\} \qquad (6.1) \\ &\leq \max\{\varphi_k'(\gamma(x_0); \gamma(x_0 + y) - \gamma(x_0)) : i \in I(x_0)\} \\ &= \varphi_{i(y)}'(\gamma(x_0); \gamma(x_0 + y) - \gamma(x_0)) \end{aligned}$$

for some $i(\boldsymbol{y}) \in I(\boldsymbol{x}_0)$. Since we assume that $\nabla\varphi_{i(\boldsymbol{y})}(\boldsymbol{z}) \geq \mathbf{0}$ contains at least one positive component for every $\boldsymbol{z} \in \mathcal{S}$ it follows from Theorem 2 of Ref. [60] that $\varphi_{i(\boldsymbol{y})}$ is *pseudoconvex* on $\mathcal{S}$ and so by (6.1) we obtain

$$f(\boldsymbol{x}_0 + \boldsymbol{y}) \geq \varphi_{i(\boldsymbol{y})}(\gamma(\boldsymbol{x}_0 + \boldsymbol{y})) \geq \varphi_{i(\boldsymbol{y})}(\gamma(\boldsymbol{x}_0)) = f(\boldsymbol{x}_0).$$

Since the above inequality holds for every $\boldsymbol{y} \in \mathbb{R}^s$ it must follow that $\boldsymbol{x}_0$ is an optimal solution of (L). □

In order to apply the ellipsoid algorithm one needs to be able to extract an element from $\mathcal{N}_{\mathcal{L}_f^<}(\boldsymbol{x}_0)$ for each $\boldsymbol{x}_0 \notin \Gamma_f$.

The following theorem follows immediately from the second part of Lemma 5.3.4 and completely characterizes $\mathcal{N}_{\mathcal{L}_f^<}(\boldsymbol{x}_0)$ in that case.

Theorem 6.2.2. *If $\boldsymbol{x}_0$ does not belong to Γ_f and $\mathcal{C}(\boldsymbol{x}_0)$ is defined as in Theorem 6.2.1 then*

$$\mathcal{N}_{\mathcal{L}_f^<}(\boldsymbol{x}_0) = \text{cone}(\mathcal{C}(\boldsymbol{x}_0))$$

As seen in Chapter 5 it is enough to know an element of $\mathcal{N}_{\mathcal{L}_f^<}(\boldsymbol{x}_0)$ to apply a central cut in $\boldsymbol{x}_0$. In order to use convex-like deep cuts as proposed in Chapter 3 one needs to restrict f to the class of lower subdifferentiable functions introduced in Definition 2.4.3.

The following result is needed for the proof of the next theorem. Although it is known (Ref. [37, 71]) we list its proof for completeness.

Lemma 6.2.2. *For every $\boldsymbol{x}, \boldsymbol{x}_0 \in r\mathcal{B}$ with $r > 0$ and $\mathcal{B}$ the open Euclidean unit ball in $\mathbb{R}^s$ and any hyperplane $\mathcal{H}$ going through $\boldsymbol{x}_0$ the orthogonal projection $\boldsymbol{x}_{\mathcal{H}}$ of $\boldsymbol{x}$ on $\mathcal{H}$ belongs to $\sqrt{2}r\mathcal{B}$.*

Proof: If $\mathcal{H}$ is a hyperplane in $\mathbb{R}^s$ going through $\boldsymbol{x}_0$ it follows that there exists some $\boldsymbol{x}^\star \in \mathbb{R}^s$ with $\|\boldsymbol{x}^\star\|_2 = 1$ such that $\mathcal{H} = \{\boldsymbol{x} \in \mathbb{R}^s : \boldsymbol{x}^{\star\top}(\boldsymbol{x} - \boldsymbol{x}_0) = 0\}$. Observe for any

$x \in r\mathcal{B}$ that its orthogonal projection $x_{\mathcal{H}}$ on $\mathcal{H}$ satisfies $x_{\mathcal{H}} = x + x^{\star^T}(x_0 - x)x^\star$ and this implies

$$\|x_{\mathcal{H}} - x^{\star^T} x_0 x^\star\|_2^2 = \|x - x^{\star^T} x x^\star\|_2^2 = \|x\|_2^2 - (x^{\star^T} x)^2 \leq \|x\|_2^2. \quad (6.2)$$

Moreover, using $x_{\mathcal{H}} \in \mathcal{H}$ we obtain by the Cauchy-Schwartz inequality that

$$\begin{aligned} \|x_{\mathcal{H}} - x^{\star^T} x_0 x^\star\|_2^2 &= \|x_{\mathcal{H}}\|_2^2 - 2x^{\star^T} x_0 x^{\star^T} x_{\mathcal{H}} + (x^{\star^T} x_0)^2 \quad (6.3) \\ &= \|x_{\mathcal{H}}\|_2^2 - (x^{\star^T} x_0)^2 \geq \|x_{\mathcal{H}}\|_2^2 - \|x_0\|_2^2. \end{aligned}$$

The desired result follows by combining inequalities (6.2) and (6.3) and using $x \in r\mathcal{B}$.

□

To prove the next result we need to introduce the following notation.

Definition 6.2.1. *The function $h : \mathbb{R}^s \longrightarrow \mathbb{R}$ differentiable on an open set $\mathcal{S}$ with $\mathbb{R}^s_+ \subseteq \mathcal{S}$ has a locally bounded gradient if for every $z \in \mathcal{S}$ there exists some $\varepsilon > 0$ and some constant M such that $\|\nabla h(y)\|_2 \leq M$ for every y belonging to $z + \varepsilon\mathcal{B}$.*

For any function $h : \mathbb{R}^s \longrightarrow \mathbb{R}$ the function $h_{\mathcal{K}}$ denotes the restriction of h to $\mathcal{K} \subseteq \mathbb{R}^s$, i.e. the domain of $h_{\mathcal{K}}$ is $\mathcal{K}$ and $h_{\mathcal{K}}$ equals h on $\mathcal{K}$.

If the gradient of the function φ_k is locally bounded it must follow for every compact set $\mathcal{K} \subseteq \mathcal{S}$ that there exists some $M > 0$ such that $\|\nabla\varphi_k(y)\|_2 \leq M$ for every $y \in \mathcal{K}$. Additionally, if $\mathcal{K}$ is also convex this implies by Theorem 9.19 of Ref. [92] that φ_k is Lipschitz continuous on $\mathcal{K}$. Since by Corollary 2.3.1 every finite-valued convex function on $\mathbb{R}^s$ is Lipschitz continuous on any compact subset it is now easy to verify that the function $f_k(x) = \varphi_k(\gamma(x))$ is also Lipschitz continuous on any compact subset of $\mathbb{R}^s$. Finally, since the Lipschitz continuity is preserved under the finite max operator we obtain that the objective function $f(x) = \max_{1 \leq k \leq \kappa} f_k(x)$ of (L) is also Lipschitz continuous on any compact subset. Knowing the Lipschitz constants of the functions φ_k and $\gamma_{\mathcal{G}_i}$ for every $1 \leq k \leq \kappa$ and $1 \leq i \leq n$ one can also easily determine the Lipschitz constant of the function f. By the above observations we obtain that the function f is quasiconvex on $\mathbb{R}^s$ and Lipschitz continuous on any compact subset.

Observe that $C(x_0)$ is defined as in Theorem 6.2.1.

Theorem 6.2.3. *If for every $1 \leq i \leq n$ the functions $\varphi_k : \mathbf{R}^n \longrightarrow \mathbf{R}$ are nondecreasing on $\mathbf{R}^n_+$ and differentiable and quasiconvex on an open convex set $\mathcal{S}$ with $\mathbf{R}^n_+ \subseteq \mathcal{S}$ and its gradients are locally bounded then the restriction $f_{\mathcal{K}}$ of the objective function $f : \mathbf{R}^s \longrightarrow \mathbf{R}$ of optimization problem (L) is boundedly lower subdifferentiable for every compact set $\mathcal{K} \subseteq \mathbf{R}^s$. Moreover, if x_0 does not belong to Γ_f it follows that* $\mathrm{cl}(\mathrm{cone}(\partial^- f_{\mathcal{K}}(x_0)))$ *equals* $\mathrm{cone}(\mathcal{C}(x_0))$ *for every compact set $\mathcal{K}$ with $x_0 \in \mathrm{int}(\mathcal{K})$ and* $\mathrm{cl}(\mathrm{cone}(\partial^- f_{\mathcal{K}}(x_0)))$ *denoting the smallest closed cone including $\partial^- f_{\mathcal{K}}(x_0)$.*

Proof: By the above remarks it follows that f is quasiconvex on $\mathbf{R}^s$ and Lipschitz continuous on any compact subset. To prove that $f_{\mathcal{K}}$ is boundedly lower subdifferentiable we will construct for each $x_0 \in \mathcal{K}$ a lower subgradient $x_0^\star$ with (uniformly) bounded norm. Observe this construction is a slight modification of a similar construction in Ref. [80]. Let $\mathcal{K}$ be some compact set and $x_0 \in \mathrm{int}(\mathcal{K})$ and suppose $\mathcal{L}^<_{f_{\mathcal{K}}}(f_{\mathcal{K}}(x_0)) = \mathcal{L}^<_f(f(x_0)) \cap \mathcal{K}$ is nonempty. Since f is quasiconvex on $\mathbf{R}^s$ and Lipschitz continuous on every compact subset it must follow that the nonempty strict lower level set $\mathcal{L}^<_f(f(x_0))$ is an open convex set. Due to x_0 not being an element of $\mathcal{L}^<_f(f(x_0))$ there exists by Theorem 2.2.9 some nonzero $u_0 \in \mathbf{R}^s$ satisfying $u_0^\top(x - x_0) < 0$ for every $x \in \mathcal{L}^<_f(f(x_0))$. Observe, if x_0 does not belong to Γ_f that by Theorem 6.2.2 every element of the cone $\mathrm{cone}(\mathcal{C}(x_0))$ satisfies the above property. Consider now the hyperplane $\mathcal{H} := \{x \in \mathbf{R}^s : u_0^\top(x - x_0) = 0\}$ and let $x_{\mathcal{H}}$ denote the orthogonal projection of x on $\mathcal{H}$. Since $\mathcal{H} \cap \mathcal{L}^<_f(f(x_0))$ is empty and $x_{\mathcal{H}}$ belongs to $\mathcal{H}$ this implies that $f(x_{\mathcal{H}}) \geq f(x_0)$ for every $x \in \mathcal{L}^<_f(f(x_0))$. Observe, since $\mathcal{K}$ is compact, that there exists some $r > 0$ satisfying $\mathcal{K} \subseteq r\mathcal{B}$ with $\mathcal{B}$ the open Euclidean unit ball. Hence by Lemma 6.2.2 and the Lipschitz continuity of f on $\sqrt{2}r\,\mathrm{cl}(\mathcal{B})$ we can find some $L > 0$ satisfying

$$f(x_0) - f(x) \leq f(x_{\mathcal{H}}) - f(x) \leq L\|x_{\mathcal{H}} - x\| \tag{6.4}$$

for every $x \in \mathcal{L}^<_f(f(x_0)) \cap \mathcal{K} \subseteq \mathcal{L}^<_f(f(x_0)) \cap r\mathcal{B}$. Moreover it follows, using $x_0 - x_{\mathcal{H}} \perp u_0$ and $x - x_{\mathcal{H}}$ parallel to u_0, that

$$\begin{aligned} u_0^\top(x - x_0) &= u_0^\top(x - x_{\mathcal{H}}) + u_0^\top(x_{\mathcal{H}} - x_0) \\ &= u_{0^\top}(x - x_{\mathcal{H}}) = \|u_0\|_2\|x - x_{\mathcal{H}}\|_2. \end{aligned} \tag{6.5}$$

Hence by (6.4) and (6.5) we obtain that $L\frac{u_0}{\|u_0\|_2}$ belongs to $\partial^- f_{\mathcal{K}}(x_0)$ and so the first part is proved. To verify the remaining part we already observed that any element of the cone $\bigcup_{\lambda>0} \lambda\mathcal{C}(\boldsymbol{x}_0)$ can be taken as $\boldsymbol{u}_0$ and this immediately implies by the above construction that $\text{cone}(\mathcal{C}(\boldsymbol{x}_0)) \subseteq \text{cl}(\text{cone}(\partial^- f_{\mathcal{K}}(\boldsymbol{x}_0)))$. Due to the definition of $\partial^- f_{\mathcal{K}}(\boldsymbol{x}_0)$ and $\boldsymbol{x}_0 \in \text{int}(\mathcal{K})$ one can easily show that $\partial^- f_{\mathcal{K}}(\boldsymbol{x}_0)$ must be a subset of the normal cone $\mathcal{N}_{\mathcal{L}_f^<}(\boldsymbol{x}_0)$ of $\mathcal{L}_f^<(f(\boldsymbol{x}_0))$ at $\boldsymbol{x}_0$ and so by Theorem 6.2.2 it follows that $\text{cl}(\text{cone}(\partial^- f_{\mathcal{K}}(\boldsymbol{x}_0))) \subseteq \text{cone}(\mathcal{C}(\boldsymbol{x}_0))$. Hence the two sets are equal and this concludes the proof. □

By a similar proof as Theorem 6.2.3 one can show that the proof of Theorem 5.5.2 holds for a quasiconvex function which is Lipschitz continuous on $\mathcal{L}_{\bar{g}}^{\leq}(0) \cap \mathcal{L}_{\bar{n}}^{\leq}(\sqrt{2}r)$.

Observe that the unconstrained location models discussed in this chapter fall into the category of the "easy" quasiconvex functions discussed in Section 5.3 and hence can be solved by Algorithm 3.13.

In the next section we present a some computational experiments.

6.3. Computational results

In order to test Algorithm 3.13 our computational experience was carried out over 300 uncorrelated planar instances of the problem taking as the disutility function of each demand point $\boldsymbol{d}_i \in \mathbb{R}^2$, $1 \leq i \leq n$, the function $\varphi_i(\boldsymbol{z}) = 100w_i \log(z_i + 1)$ for $1 \leq i \leq \lceil n/3 \rceil$, $\varphi_i(\boldsymbol{z}) = 100w_i \arctan(z_i)$ for $\lceil n/3 \rceil + 1 \leq i \leq \lceil 2n/3 \rceil$ and $\varphi_i(\boldsymbol{z}) = 5w_i z_i$ for $\lceil 2n/3 \rceil + 1 \leq i \leq n$ with $\lceil x \rceil$ denoting the ceiling of $x \in \mathbb{R}$. Assigning to the nonlinear (concave) disutility functions a bigger weight than to the linear ones prevents that the linear functions dominate the concave functions prohibiting f to become convex.

It is easy to verify, due to $\|\boldsymbol{x}\|_p \leq \|\boldsymbol{x}\|_1 \leq \sqrt{s}\|\boldsymbol{x}\|_2$ for every $1 \leq p \leq \infty$ and $\boldsymbol{x} \in \mathbb{R}^s$, that for $s = 2$ the constant

$$\sqrt{2}\max\left\{\max_{1\leq i\leq\lceil 2n/3\rceil}\{100w_i\}, \max_{\lceil 2n/3\rceil+1\leq i\leq n}\{5w_i\}\right\}$$

is a Lipschitz constant for the function f.

The instances were randomly generated in the following way. We start by describing the selection of the problem parameters.

The number n of demand points belongs to $\{5, 25, 50, 100, 250\}$.

For the ℓ_p-norm being used, we take $p_i \in \{1.1, 1.5, 1.9, 2.1, 3.0\}$ and construct two different classes of examples. For the first class one value is assigned to all the demand points while for the second class, denoted in the tables as "mix", to each demand point a value p_i randomly selected from the set $\{1.1, 1.5, 1.9, 2.1, 3.0\}$ is assigned. The weight w_i of demand point $\boldsymbol{d}_i$ is determined as follows. We uniformly draw numbers from the interval $[0, 1]$, say $\widetilde{w}_i$, $1 \leq i \leq n$, and set w_i equal to $\widetilde{w}_i / \sum_{j=1}^{n} \widetilde{w}_j$ for every $1 \leq i \leq n$.

Now we describe the procedure to generate the demand points. All the demand points are generated within the square $[0, 250] \times [0, 250]$, for which a clustered structure is created using the following procedure. First we draw two integers m_1 and m_2 ranging from 1 to 20, and then we divide the square $[0, 250] \times [0, 250]$ into $(m_1 + 1)(m_2 + 1)$ subsquares by generating randomly m_1 x-axis coordinates and m_2 y-axis coordinates in $]0, 250[$ (see Figure 6.1). Then we label these subsquares from 1 to $(m_1 + 1)(m_2 + 1)$.

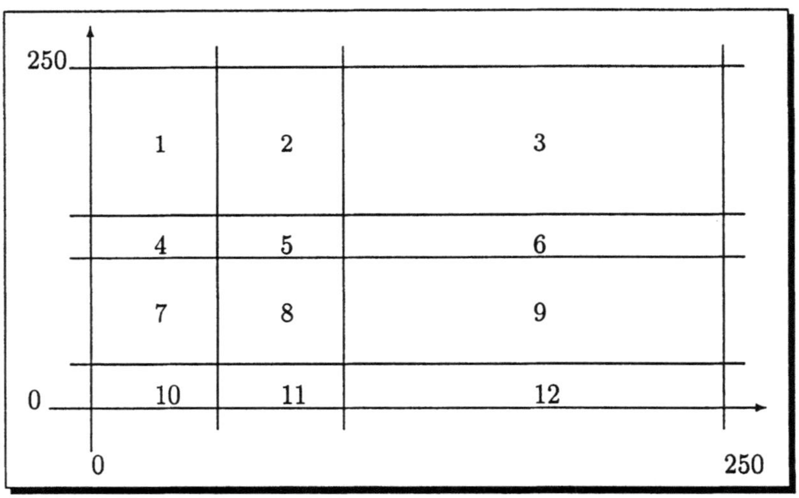

Figure 6.1: Generating clustered data

Subsequently we randomly choose according to these labels some given number of subsquares. In each chosen subsquare we uniformly draw a given number of demand points. Finally, the remaining demand points are uniformly drawn from the original square $[0, 250] \times [0, 250]$ and added to the already existing set of demand points, in a total of n points.

Finally, the execution is stopped with a relative error (Section 3.3) less than 5×10^{-6} or 5×10^{-11}. An optimality test as described in Section 4.6 is also implemented.

For each pair (n, p), 10 uncorrelated instances of the problem were generated according to the procedure described above and each of them was solved by the central cut and the deep cut versions of the algorithm. Also for each value of n, 10 uncorrelated instances were generated with mixed norms and also solved by both versions of the algorithm.

The results are summarized in Table 6.1 and Table 6.2 where each row corresponds to averages of 10 instances.

Under *central cuts* we list under *time* the average time taken by the central cut version of the algorithm in seconds of the Compaq Prolinea, and under *it* the average number of iterations. All these iterations correspond to objective cuts since for these instances no norm cut was generated.

Under *deep cuts* we list under *time* the average time taken by the deep cut version of the algorithm in seconds of the Compaq Prolinea, the average total number of iterations (*it*), the average depth of an objective cut (α). Finally under *% d* the percentage of objective cuts that were deep is listed and the corresponding average depth (average taken within the deep objective cuts only) under α_d.

The last pair of columns, under *% reduction*, *time* corresponds to the percentage by which the application of deep cuts reduces the total time and *it* has a similar meaning for the iterations.

Notice that the seemingly small average depth may be explained by the dependence between the value of the Lipschitz constant and the depth of the cut.

For the convex case more encouraging results are reported in Ref. [38] with an average depth of cuts ten times bigger and about 16% reduction in iterations and time.

In order to report some results on convex functions we generated similar examples where the disutility function of every demand point is taken linear, i.e. $\varphi_i(z) = w_i z_i$.

problem		*central cuts*		*deep cuts*					*% reduction*	
n	p	*time*	*it*	*time*	*it*	α	% d	α_d	*time*	*it*
5	1.1	0.059	90.6	0.058	89.6	0.0049	70.3	0.0070	0.6	1.1
5	1.5	0.064	98.1	0.064	97.0	0.0035	69.3	0.0052	0.5	1.1
5	1.9	0.063	97.0	0.063	96.3	0.0030	72.8	0.0041	0.1	0.7
5	2.1	0.061	92.6	0.060	91.6	0.0038	71.2	0.0053	0.4	1.1
5	3.0	0.064	97.9	0.063	96.5	0.0046	71.5	0.0065	0.8	1.4
5	mix	0.065	100.2	0.065	99.3	0.0031	69.9	0.0044	0.3	0.9
25	1.1	0.221	97.5	0.218	96.0	0.0034	71.4	0.0047	1.3	1.5
25	1.5	0.230	101.7	0.229	100.9	0.0033	74.3	0.0045	0.6	0.8
25	1.9	0.225	99.2	0.223	98.1	0.0028	75.7	0.0036	0.9	1.1
25	2.1	0.224	98.9	0.221	97.5	0.0026	73.3	0.0035	1.2	1.4
25	3.0	0.247	109.2	0.246	108.4	0.0027	74.2	0.0036	0.5	0.7
25	mix	0.240	106.0	0.239	105.4	0.0031	72.0	0.0043	0.3	0.6
50	1.1	0.446	103.4	0.438	101.2	0.0045	77.0	0.0058	1.8	2.1
50	1.5	0.441	102.2	0.438	101.2	0.0032	75.7	0.0042	0.6	1.0
50	1.9	0.482	111.8	0.478	110.3	0.0033	75.5	0.0044	1.0	1.3
50	2.1	0.468	108.5	0.465	107.4	0.0034	75.7	0.0044	0.7	1.0
50	3.0	0.439	101.8	0.440	101.6	0.0028	73.7	0.0037	-0.1	0.2
50	mix	0.481	111.6	0.478	110.4	0.0032	72.6	0.0043	0.7	1.1
100	1.1	0.789	93.8	0.777	92.2	0.0043	74.6	0.0057	1.5	1.7
100	1.5	0.872	103.6	0.859	101.8	0.0033	74.1	0.0045	1.6	1.7
100	1.9	0.920	109.3	0.914	108.5	0.0035	75.9	0.0046	0.6	0.7
100	2.1	0.901	107.0	0.882	104.6	0.0038	76.1	0.0049	2.1	2.2
100	3.0	0.901	107.0	0.893	105.9	0.0030	75.4	0.0039	0.9	1.0
100	mix	0.873	103.7	0.868	103.0	0.0035	74.3	0.0047	0.5	0.7
250	1.1	2.110	102.2	2.066	100.0	0.0047	75.2	0.0063	2.1	2.2
250	1.5	2.205	106.7	2.181	105.5	0.0040	75.6	0.0053	1.1	1.1
250	1.9	2.297	111.1	2.267	109.6	0.0039	76.0	0.0051	1.3	1.4
250	2.1	2.358	114.1	2.324	112.4	0.0038	77.9	0.0049	1.4	1.5
250	3.0	2.297	111.1	2.272	109.8	0.0036	78.2	0.0046	1.1	1.2
250	mix	2.386	115.5	2.348	113.6	0.0045	75.8	0.0060	1.6	1.6

Table 6.1: Results for quasiconvex location problems, $\varepsilon = 5 \times 10^{-6}$

problem		*central cuts*		*deep cuts*					*% reduction*	
n	p	*time*	*it*	*time*	*it*	α	% d	α_d	*time*	*it*
5	1.1	0.101	156.8	0.101	154.7	0.0049	71.5	0.0068	0.7	1.3
5	1.5	0.108	165.8	0.108	164.2	0.0035	68.9	0.0052	0.3	1.0
5	1.9	0.109	167.3	0.109	165.5	0.0030	73.5	0.0041	0.5	1.1
5	2.1	0.103	158.2	0.103	156.1	0.0037	70.3	0.0053	0.7	1.3
5	3.0	0.108	166.1	0.107	163.4	0.0046	70.3	0.0067	1.0	1.6
5	mix	0.111	170.3	0.110	168.0	0.0032	70.5	0.0045	0.7	1.4
25	1.1	0.395	174.2	0.390	171.6	0.0036	72.9	0.0050	1.3	1.5
25	1.5	0.399	176.2	0.396	174.2	0.0033	73.1	0.0045	0.9	1.1
25	1.9	0.388	171.0	0.387	170.5	0.0026	74.4	0.0035	0.0	0.3
25	2.1	0.384	169.6	0.380	167.5	0.0025	71.6	0.0035	1.0	1.2
25	3.0	0.424	187.2	0.423	186.0	0.0027	73.3	0.0037	0.4	0.6
25	mix	0.419	184.7	0.417	183.4	0.0031	71.7	0.0043	0.4	0.7
50	1.1	0.787	182.4	0.777	179.5	0.0043	75.1	0.0058	1.2	1.6
50	1.5	0.763	176.9	0.756	174.6	0.0033	74.6	0.0044	0.9	1.3
50	1.9	0.818	189.6	0.819	189.2	0.0033	73.5	0.0045	-0.2	0.2
50	2.1	0.808	187.2	0.801	184.9	0.0034	74.5	0.0045	0.9	1.2
50	3.0	0.758	175.7	0.752	173.7	0.0027	71.8	0.0037	0.8	1.1
50	mix	0.827	191.8	0.821	189.8	0.0033	73.0	0.0044	0.7	1.0
100	1.1	1.402	166.6	1.375	163.1	0.0041	72.6	0.0057	1.9	2.1
100	1.5	1.501	178.2	1.491	176.7	0.0034	72.3	0.0048	0.7	0.8
100	1.9	1.571	186.6	1.500	177.9	0.0036	74.9	0.0047	4.5	4.7
100	2.1	1.513	179.7	1.507	178.7	0.0038	75.2	0.0050	0.4	0.6
100	3.0	1.525	181.1	1.493	177.0	0.0029	73.4	0.0040	2.1	2.3
100	mix	1.508	179.2	1.498	177.6	0.0035	73.1	0.0048	0.7	0.9
250	1.1	3.742	181.2	3.649	176.6	0.0049	74.3	0.0066	2.5	2.5
250	1.5	3.809	184.3	3.720	179.9	0.0041	73.5	0.0056	2.3	2.4
250	1.9	3.894	188.3	3.782	182.8	0.0040	74.3	0.0053	2.9	2.9
250	2.1	3.927	190.0	3.954	191.2	0.0038	74.6	0.0051	-0.7	-0.6
250	3.0	3.770	182.3	3.770	182.2	0.0036	75.5	0.0048	0.0	0.1
250	mix	4.040	195.5	3.947	190.9	0.0043	72.9	0.0059	2.3	2.4

Table 6.2: Results for quasiconvex location problems, $\varepsilon = 5 \times 10^{-11}$

This is known as the Rawls problem and the corresponding results are given in Table 6.3 and Table 6.4.

For this problem the influence of deep cuts is much more encouraging.

problem		*central cuts*		*deep cuts*					*% reduction*	
n	p	*time*	*it*	*time*	*it*	α	*% d*	α_d	*time*	*it*
5	1.1	0.041	68.8	0.030	49.7	0.0891	62.2	0.1422	27.1	27.8
5	1.5	0.046	76.7	0.037	61.0	0.0676	58.2	0.1114	19.5	20.5
5	1.9	0.047	77.9	0.032	53.4	0.0968	64.4	0.1492	30.7	31.5
5	2.1	0.044	73.6	0.033	54.4	0.0810	60.5	0.1311	25.3	26.1
5	3.0	0.045	75.3	0.034	56.0	0.0799	60.9	0.1286	24.8	25.6
5	mix	0.050	84.1	0.040	66.5	0.0677	54.6	0.1110	20.1	20.9
25	1.1	0.159	75.5	0.111	52.4	0.0942	65.8	0.1433	30.3	30.6
25	1.5	0.184	87.6	0.144	68.1	0.0633	57.3	0.1071	21.9	22.3
25	1.9	0.162	77.0	0.113	53.6	0.0979	61.9	0.1509	30.1	30.4
25	2.1	0.171	81.3	0.127	60.2	0.0796	60.1	0.1249	25.6	26.0
25	3.0	0.170	80.8	0.130	61.6	0.0741	57.6	0.1228	23.4	23.8
25	mix	0.166	78.8	0.118	55.8	0.0889	63.3	0.1358	28.8	29.2
50	1.1	0.309	77.0	0.213	52.8	0.0952	64.2	0.1478	31.2	31.4
50	1.5	0.323	80.4	0.233	57.9	0.0868	63.0	0.1329	27.7	28.0
50	1.9	0.324	80.5	0.246	61.0	0.0788	61.1	0.1199	23.9	24.2
50	2.1	0.325	81.0	0.255	63.2	0.0674	59.3	0.1065	21.7	22.0
50	3.0	0.300	74.6	0.215	53.3	0.0855	62.9	0.1350	28.3	28.6
50	mix	0.334	83.0	0.244	60.6	0.0830	61.1	0.1287	26.7	27.0
100	1.1	0.611	77.7	0.421	53.4	0.0958	65.4	0.1463	31.1	31.3
100	1.5	0.635	80.8	0.458	58.1	0.0867	62.3	0.1369	28.0	28.1
100	1.9	0.653	83.0	0.523	66.4	0.0662	56.3	0.1105	19.9	20.0
100	2.1	0.616	78.4	0.481	61.0	0.0731	59.2	0.1167	22.0	22.2
100	3.0	0.672	85.5	0.563	71.5	0.0525	53.0	0.0928	16.2	16.4
100	mix	0.616	78.4	0.473	60.0	0.0749	59.5	0.1208	23.3	23.5
250	1.1	1.528	79.2	1.054	54.6	0.0935	63.2	0.1476	31.0	31.1
250	1.5	1.562	80.9	1.090	56.4	0.0954	64.5	0.1445	30.2	30.3
250	1.9	1.667	86.3	1.225	63.4	0.0796	61.0	0.1255	26.5	26.5
250	2.1	1.706	88.4	1.333	69.0	0.0683	56.4	0.1134	21.9	21.9
250	3.0	1.701	88.1	1.309	67.8	0.0693	57.4	0.1162	23.0	23.0
250	mix	1.649	85.4	1.156	59.8	0.0917	63.2	0.1410	29.9	30.0

Table 6.3: Results for convex location problems, $\varepsilon = 5 \times 10^{-6}$

problem		*central cuts*		*deep cuts*					*% reduction*	
n	p	*time*	*it*	*time*	*it*	α	% d	α_d	*time*	*it*
5	1.1	0.082	137.6	0.057	95.4	0.0939	65.0	0.1440	30.0	30.7
5	1.5	0.089	150.0	0.071	117.7	0.0674	59.2	0.1095	20.7	21.5
5	1.9	0.091	151.8	0.061	101.6	0.1015	66.4	0.1503	32.3	33.1
5	2.1	0.086	144.7	0.063	104.8	0.0836	63.0	0.1302	26.8	27.6
5	3.0	0.087	145.5	0.063	103.8	0.0867	63.3	0.1351	27.9	28.7
5	mix	0.096	161.9	0.077	127.4	0.0667	53.8	0.1090	20.4	21.3
25	1.1	0.320	152.3	0.213	100.7	0.1050	68.4	0.1529	33.6	33.9
25	1.5	0.358	170.2	0.287	136.2	0.0595	54.3	0.1025	19.6	20.0
25	1.9	0.315	150.0	0.220	104.3	0.0971	63.1	0.1456	30.1	30.5
25	2.1	0.331	157.6	0.244	115.5	0.0838	61.1	0.1288	26.4	26.7
25	3.0	0.326	155.2	0.250	118.5	0.0740	58.6	0.1199	23.3	23.6
25	mix	0.327	155.8	0.225	106.8	0.0969	65.5	0.1430	31.1	31.5
50	1.1	0.615	153.1	0.418	103.7	0.0996	66.5	0.1493	32.0	32.3
50	1.5	0.630	157.0	0.449	111.4	0.0896	64.4	0.1347	28.8	29.0
50	1.9	0.630	156.8	0.482	119.4	0.0766	60.1	0.1182	23.6	23.9
50	2.1	0.634	157.8	0.501	124.3	0.0666	58.7	0.1050	20.9	21.2
50	3.0	0.580	144.4	0.404	100.2	0.0951	65.7	0.1441	30.4	30.6
50	mix	0.650	161.7	0.462	114.5	0.0934	63.5	0.1369	28.9	29.2
100	1.1	1.234	157.0	0.815	103.5	0.1075	68.8	0.1561	34.0	34.1
100	1.5	1.241	157.8	0.873	110.8	0.0916	63.8	0.1389	29.6	29.8
100	1.9	1.284	163.2	0.998	126.6	0.0722	58.0	0.1148	22.3	22.4
100	2.1	1.188	151.0	0.915	116.1	0.0753	59.2	0.1189	23.0	23.1
100	3.0	1.299	165.1	1.093	138.7	0.0502	52.3	0.0877	15.8	16.0
100	mix	1.208	153.7	0.899	114.2	0.0812	61.6	0.1250	25.6	25.7
250	1.1	3.100	160.7	2.046	106.0	0.1049	67.4	0.1547	34.0	34.0
250	1.5	3.041	157.5	2.044	105.8	0.1064	67.3	0.1531	32.8	32.8
250	1.9	3.274	169.5	2.335	120.8	0.0892	63.8	0.1327	28.7	28.7
250	2.1	3.299	170.9	2.573	133.2	0.0698	56.2	0.1134	22.0	22.1
250	3.0	3.316	171.8	2.544	131.7	0.0704	56.3	0.1160	23.3	23.3
250	mix	3.260	168.8	2.222	115.0	0.0988	65.6	0.1456	31.8	31.9

Table 6.4: Results for convex location problems, $\varepsilon = 5 \times 10^{-11}$

VII Conclusions

In this book we introduce an improved version of the ellipsoid method for convex and quasiconvex programs. This improved version uses so-called norm cuts.

Convergence rates for this improved ellipsoid method applied to convex programs are presented and the proofs of these results are simple and elementary. Moreover, since we also construct an implementable separation oracle for the strict lower level sets of quasiconvex functions we were able to derive a variant of the improved version of the ellipsoid method capable of solving problems with a continuous quasiconvex objective function and convex constraints. These problems are called quasiconvex programs. Although the analysis of the separation oracle for the quasiconvex objective function is more complex it is still possible to give a clear and direct interpretation of the main theoretical results. Also for this new method similar rate of convergence results were proved as for the convex case and this appears to be the first application of a version of the ellipsoid method to quasiconvex programs with a convergence proof.

The technique proposed in this book to construct a separation oracle for quasiconvex functions can also, in principle, be used by other procedures which rely on a separation oracle. Moreover, besides norm cuts we also introduce so-called box and parallel cuts that can be embedded in the ellipsoid method.

We like to remark that this research was motivated by the need to solve general quasiconvex location models and since these models can now be solved this may contribute to enlarge the range of possible real life applications of continuous location. Our preliminary computational experience shows that the ellipsoid method is a useful method in convex and quasiconvex continuous location. Hence, it is interesting to test the behavior of our methods to other members of the above class.

Bibliography

[1] M. ANDRAMONOV, *Minimizing quasi-convex functions with the use of marginal functions*, in Proceedings of the workshop on Generalized Convexity and its Applications, Milan, 1994.

[2] ——, *Quasi-convexity of marginal functions*, tech. rep., University of Venice, Italy, 1994.

[3] M. ANDRAMONOV AND A. ELLERO, *A quasi-convexity condition for marginal functions and its applications*, in Proceedings of the workshop on Scalar and Vector Optimization Problems and their applications in Economics, Milan, 1995.

[4] K. M. ANSTREICHER, *A monotonic projective algorithm for fractional linear programming*, Algorithmica, 1 (1986), pp. 483–498.

[5] ——, *On Vaidya's volumetric cutting plane method for convex programming*, tech. rep., University of Iowa, Iowa City, IA 52242, USA, 1994.

[6] D. S. ATKINSON AND P. V. VAIDYA, *An analytic center based cutting plane algorithm for convex programming*, tech. rep., University of Illinois at Urbana-Champaign, Illinois, USA, 1992.

[7] M. AVRIEL, W. E. DIEWERT, S. SCHAIBLE, AND I. ZANG, *Generalized Concavity*, Plenum Press, New York, 1988.

[8] W. BARITOMPA, *Multidimensional bisection: A dual viewpoint*, Computers and Mathematics with Applications, 27 (1994), pp. 11–22.

[9] A. BARROS AND J. FRENK, *Generalized fractional programming and cutting plane algorithms*, Journal of Optimization Theory and Applications, 87 (1995), pp. 103–120.

[10] A. I. M. B. BARROS, *Discrete and Fractional Programming Techniques for Location Models*, no. 89 in Tinbergen Institute Research Series, Thesis Publishers, Amsterdam, The Netherlands, January 1995.

[11] M. S. BAZARAA, H. D. SHERALI, AND C. M. SHETTY, *Nonlinear Programming: Theory and Algorithms*, Wiley, New York, second ed., 1993.

[12] M. S. BAZARAA AND C. M. SHETTY, *Nonlinear Programming: Theory and Algorithms*, Wiley, New York, 1979.

[13] W. BERENS AND F. J. KOERLING, *Estimated road distances by mathematical functions*, European Journal of Operational Research, 21 (1972), pp. 54–56.

[14] R. G. BLAND, D. GOLDFARB, AND M. J. TODD, *The ellipsoid method: A survey*, Operations Research, 29 (1981), pp. 1039–1091.

[15] A. CAMBINI, E. CASTAGNOLI, L. MARTEIN, P. MAZZOLENI, AND S. SCHAIBLE, eds., *Generalized Convexity and Fractional Programming With Economic Applications*, vol. 345, Berlin, 1990, Springer-Verlag.

[16] E. CARRIZOSA AND F. PLASTRIA, *Locating an undesirable facility by generalized cutting planes*, tech. rep., Universitad de Sevilla, C/Trafia s/n, 41012, Sevilla, Spain, 1995.

[17] A. CHARNES AND W. W. COOPER, *Programming with linear fractionals*, Naval Research Logistics Quarterly, 9 (1962), pp. 181–186.

[18] E. W. CHENEY AND A. A. GOLDSTEIN, *A Newton's method for convex programming and Tchebycheff approximation*, Numeric Mathematics, 1 (1959), pp. 253–268.

[19] F. H. CLARKE, *Optimization and Nonsmooth Analysis*, John Wiley, New York, 1983.

[20] B. D. CRAVEN, *Fractional Programming*, Heldermann-Verlag, Berlin, 1988.

[21] J. P. CROUZEIX, *Continuity and differentiability properties of quasiconvex functions on* $\mathbb{R}^n$, in Generalized Concavity in Optimization and Economics, S. Schaible and W. T. Ziemba, eds., New York, 1981, Academic Press, pp. 109–130.

[22] ——, *Some Differentiability Properties of Quasiconvex Functions on* $\mathbb{R}^n$, vol. 30 of Lecture Notes in Control & Information Sciences, Springer-Verlag, New York, 1981, pp. 9–20.

[23] ——, *About differentiability of order one of quasiconvex functions on* $\mathbb{R}^n$, Journal of Optimization Theory and Applications, 36 (1982), pp. 367–385.

[24] G. Dahlquist, Å. Björck, and N. Anderson, *Numerical Methods*, Prentice-Hall, Englewoods Cliffs, N.J., 1974.

[25] V. F. Dem'yanov and L. C. W. Dixon, eds., *Quasidifferential Calculus*, vol. 29, Mathematical Programming Study, North-Holland, 1986.

[26] D. den Hertog, *Interior Point Approach to Linear, Quadratic and Convex Programming - Algorithms and Complexity*, vol. 277 of Mathematics and Applications, Kluwer Academic Publisher, 1994.

[27] D. den Hertog, J. Kaliski, C. Roos, and T. Terlaky, *A logarithmic barrier cutting plane method for convex programming*, tech. rep., Delft University of Technology, Delft, The Netherlands, 1994. Revised version submitted to Annals of Operations Research.

[28] M. E. Dyer, *Linear time algorithms for two and three variable linear programs*, SIAM Journal on Computing, 13 (1984), pp. 31–45.

[29] S. T. Dziuban, J. G. Ecker, and M. Kupferschmid, *Using deep cuts in an ellipsoid algorithm for nonlinear programming*, Mathematical Programming Study, 25 (1985), pp. 93–107.

[30] J. G. Ecker and M. Kupferschmid, *An ellipsoid algorithm for nonlinear programming*, Mathematical Programming, 27 (1983), pp. 83–106.

[31] ——, *A computational comparison of the ellipsoid algorithm with several nonlinear programming algorithms*, SIAM Journal on Control and Optimization, 23 (1985), pp. 657–674.

[32] W. Fenchel, *A remark on convex sets and polarity*, in Communication Seminar on Mathematics, University of Lund Supplementary Volume, Lund, 1952, University of Lund, pp. 22–89.

[33] L. R. FOULDS AND H. W. HAMACHER, *Optimal bin location and sequencing in printed circuit board assembly*, Tech. Rep. 181, University of Kaiserslauteren, Germany, 1990.

[34] J. B. G. FRENK AND J. GROMICHO, *An elementary rate of convergence proof for the deep cut ellipsoid algorithm*, in Recent advances in nonsmooth optimization, D.-Z. Du, L. Qi, and R. S. Womersley, eds., World Scientific Publishers, Singapore, 1995, pp. 106–120.

[35] J. B. G. FRENK, J. GROMICHO, AND D. M. L. DIAS, *Duality theory for convex/quasiconvex functions and its application to optimization: An overview*, in Generalized Convexity, S. Komlósi, T. Rapcsák, and S. Schaible, eds., Springer Verlag, 1994, pp. 153–170.

[36] J. B. G. FRENK, J. GROMICHO, M. FIDRICH, AND S. ZHANG, *An efficient algorithm to check whether* 0 *belongs to the convex hull of a finite number of* ℓ_p*-circles*, Tech. Rep. 9204/A, Econometric Institute, Erasmus University, Rotterdam, The Netherlands, 1992. 24 pages.

[37] J. B. G. FRENK, J. GROMICHO, F. PLASTRIA, AND S. ZHANG, *A deep cut ellipsoid algorithm and quasiconvex programming*, in Generalized Convexity, S. Komlósi, T. Rapcsák, and S. Schaible, eds., Springer Verlag, 1994, pp. 62–76.

[38] J. B. G. FRENK, J. GROMICHO, AND S. ZHANG, *A deep cut ellipsoid algorithm for convex programming: Theory and applications*, Mathematical Programming, 63 (1994), pp. 83–108.

[39] ——, *Quasiconvex functions: How to separate, if you must!*, Tech. Rep. TI 94-59, Tinbergen Institute, Rotterdam, The Netherlands, 1994. Submitted.

[40] ——, *General models in min-max continuous location: Theory and solution techniques*, Journal of Optimization Theory and Applications, 89 (1996, April), pp. 39–63.

[41] ——, *General models in min-max planar location: Checking optimality conditions*, Journal of Optimization Theory and Applications, 89 (1996, April), pp. 65–87.

[42] R. W. FREUND AND F. JARRE, *An interior-point method for fractional programs with convex constraints*, Mathematical Programming, 67 (1994), pp. 407–440.

[43] ——, *An interior-point method for multifractional programs with convex constraints*, Journal of Optimization Theory and Applications, 85 (1995), pp. 125–161.

[44] P. E. GILL, G. H. GOLUB, W. MURRAY, AND M. A. SAUNDERS, *Methods for modifying matrix factorizations*, Mathematics of Computation, 28 (1974), pp. 505–535.

[45] P. E. GILL, W. MURRAY, AND M. A. SAUNDERS, *Methods for computing and modifying the LDV factors of a matrix*, Mathematics of Computation, 29 (1975), pp. 1051–1077.

[46] J.-L. GOFFIN, *Convergence rates of the ellipsoid method on general convex functions*, Mathematics of Operations Research, 8 (1983), pp. 135–150.

[47] J.-L. GOFFIN, Z.-Q. LUO, AND Y. YE, *Complexity analysis of an interior cutting plane method for convex feasibility problems*, SIAM Journal of Optimization, 63 (1996), pp. 638–652.

[48] D. GOLDFARB AND M. J. TODD, *Modifications and implementation of the ellipsoid algorithm for linear programming*, Mathematical Programming, 23 (1982), pp. 1–19.

[49] G. H. GOLUB AND C. F. VAN LOAN, *Matrix Computations*, John Hopkins Unviversity Press, Baltimore, 1983.

[50] J. A. S. GROMICHO, *Quasiconvex Optimization and Location Theory*, no. 90 in Tinbergen Institute Research Series, Thesis Publishers, Amsterdam, The Netherlands, January 1995.

[51] M. GRÖTSCHEL, L. LOVÁSZ, AND A. SCHRIJVER, *Geometric Algorithms and Combinatorial Optimization*, Springer-Verlag, Berlin Heidelberg, 1988.

[52] J.-B. HIRIART-URRUTY AND C. LEMARÉCHAL, *Convex Analysis and Minimization Algorithms I: Fundamentals*, vol. 305 of A Series of Comprehensive Studies in Mathematics, Springer-Verlag, Berlin, 1993.

[53] ——, *Convex Analysis and Minimization Algorithms II: Advanced Theory and Bundle Methods*, vol. 306 of A Series of Comprehensive Studies in Mathematics, Springer-Verlag, Berlin, 1993.

[54] R. Horst and P. Pardalos, eds., *Handbook of Global Optimization*, vol. 2 of Nonconvex Optimization and Its Applications, Kluwer Academic Publishers, Dordrecht, 1995.

[55] H. Idrissi, P. Loridan, and C. Michelot, *Approximation of solutions for location problems*, Journal of Optimization Theory and Applications, 56 (1988), pp. 127–143.

[56] B. Jansen, C. Roos, and T. Terlaky, *On Vaidya's volumetric center method for convex programming*, in Volume of extended abstracts, I. Maros, ed., APMOD 1993, Budapest, Hungary, 1993.

[57] M. Kallay, *Convex hulls made easy*, Information Processing Letters, 22 (1986), p. 161.

[58] J. E. Kelley, *The cutting plane method for solving convex problems*, Journal of the Society for the Industrial and Applied Mathematics, 8 (1960), pp. 703–712.

[59] L. G. Khachiyan, *A polynomial algorithm in linear programming (in Russian)*, Doklady Akademii Nauk SSSR, 244 (1979), pp. 1093–1979.

[60] S. Komlósi, *Some properties of nondifferentiable pseudoconvex functions*, Mathematical Programming, 26 (1983), pp. 232–237.

[61] S. Komlósi, T. Rapcsák, and S. Schaible, eds., *Generalized Convexity*, Springer Verlag, 1994.

[62] I. V. Konnov, *Methods of Nondifferentiable Optimization*, Kazan University Publishing Office, Kazan, Russia, 1993. in Russian.

[63] ——, *Aplications of the combined relaxation method to finding equilibrium points of a quasi-convex-concave function*, Izvestiya VUZ. Matematika, (1994).

[64] P. Lancaster and M. Tismenetsky, *The Theory of Matrices*, Academic Press, New York, second ed., 1985.

[65] A. Y. LEVIN, *On an algorithm for the minimization of convex functions*, Doklady Akademii Nauk SSSR, 160 (1965), pp. 1244–1247. in Russian.

[66] R. F. LOVE AND J. G. MORRIS, *Modelling intercity road distances by mathematical functions*, Operational Research Quarterly, 23 (1972), pp. 61–71.

[67] R. F. LOVE, J. G. MORRIS, AND G. O. WESOLOWSKY, *Facility Location: Models & Methods*, Elsevier Science Publisher B.V., North-Holland, 1988.

[68] D. G. LUENBERGER, *Linear and Nonlinear Programming*, Addison-Wesley, Reading Massachusetts, 1984.

[69] H. J. LUTHI, *On the solution of variational inequalities by the ellipsoid method*, Mathematics of Operations Research, 10 (1985), pp. 515–522.

[70] J. T. MARTI, *Konvexe Analysis*, Birkhauser Verlag, Basel, 1977.

[71] J. E. MARTINEZ-LEGAZ, *On lower subdifferentiable functions*, in Trends in Mathematical Optimization, K. H. Hoffman, J.-B. Hiriart-Urruty, C. Lemaréchal, and J. Zowe, eds., Birkhäuser-Verlag, Basel, 1988.

[72] N. MEGIDDO, *Linear programming in linear time when the dimension is fixed*, Journal of the Association for Computing Machinery, 31 (1984), pp. 114–127.

[73] C. MICHELOT, *The mathematics of continuous location*, Studies in Locational Analysis, 5 - ISOLDE VI survey papers (1993), pp. 59–83.

[74] G. L. NEMHAUSER AND L. A. WOLSEY, *Integer and Combinatorial Optimization*, Wiley, New York, 1988.

[75] A. S. NEMIROVSKY AND D. B. YUDIN, *Problem Complexity and Method Efficiency in Optimization*, John Wiley & Sons, Chichester, 1983.

[76] J. M. ORTEGA AND W. C. RHEINBOLDT, *Iterative Solutions of Nonlinear Equations in Several Variables*, Academic Press, New York, 1970.

[77] U. PASSY AND E. Z. PRISMAN, *A convex-like duality scheme for quasiconvex programs*, Mathematical Programming, 32 (1985), pp. 278–300.

[78] J. P. PENOT AND M. VOLLE, *On quasiconvex duality*, Mathematics of Operations Research, 15 (1990), pp. 597–625.

[79] F. PLASTRIA, *Localization in single facility location*, European Journal of Operational Research, 18 (1984), pp. 215–219.

[80] ——, *Lower subdifferentiable functions and their minimization by cutting planes*, Journal of Optimization Theory and Applications, 46 (1985), pp. 37–53.

[81] ——, *The minimization of lower subdifferential functions under nonlinear constraints: An all feasible cutting plane algorithm*, Journal of Optimization Theory and Applications, 57 (1988), pp. 463–484.

[82] ——, *On destination optimality in asymmetric distance Fermat-Weber problems*, Annals of Operations Research, 40 (1992), pp. 355–369.

[83] ——, *Continuous location anno 1992: A progress report*, Studies in Locational Analysis - ISOLDE VI Survey Papers, 5 (1993), pp. 85–127.

[84] B. T. POLJAK, *A general method for solving extremum problems*, Soviet Mathematics Doklady, 8 (1967), pp. 593–597.

[85] J. PONSTEIN, *Seven kinds of convexity*, SIAM Review, 9 (1967), pp. 115–119.

[86] F. P. PREPARATA AND M. I. SHAMOS, *Computational Geometry: An Introduction*, Springer-Verlag, New York, 1985.

[87] B. N. PSHENICHNYI, *Necessary Conditions for an Extremum*, Marcel Dekker, New York, 1971.

[88] L. QI, A. RUSZCZYŃSKI, AND R. WOMERSLEY, eds., *Computational Nonsmooth Optimization*, vol. 76 of Mathematical Programming series B, North-Holland, Amsterdam, The Netherlands, March 1997.

[89] R. T. ROCKAFELLAR, *Convex Analysis*, Princeton University Press, Princeton, New Jersey, 1972.

[90] ——, *Conjugate Duality and Optimization*, SIAM, Philadelphia, 1974.

[91] I. ROSENHOLTZ, *There is no differentiable metric on* $\mathbb{R}^n$, American Mathematical Monthly, 86 (1979), pp. 585–586.

[92] W. RUDIN, *Principles of Mathematical Analysis*, McGraw-Hill, Auckland, third ed., 1976.

[93] S. SCHAIBLE AND W. T. ZIEMBA, eds., *Generalized Concavity in Optimization and Economics*, New York, 1981, Academic Press.

[94] A. SCHRIJVER, *Theory of Linear and Integer Programming*, Wiley, Chichester, 1986.

[95] R. SEIDEL, *Small-dimensional linear programming and convex hulls made easy*, Discrete & Computational Geometry, 6 (1991), pp. 423–434.

[96] N. Z. SHOR, *Convergence rate of the gradient descent method with dilation of the space*, Cybernetics, 6 (1970), pp. 102–108.

[97] ——, *Utilization of the operation of space dilation in the minimization of convex functions*, Cybernetics, 6 (1970), pp. 7–15.

[98] ——, *Cut-off method with space extension in convex programming problems*, Cybernetics, 13 (1977), pp. 94–96.

[99] ——, *New development trends in nondifferentiable optimization*, Cybernetics, 13 (1977), pp. 881–886.

[100] ——, *Minimization Methods for Non-Differentiable Functions*, Springer Series in Computational Mathematics, Springer-Verlag, Berlin, 1985.

[101] N. Z. SHOR AND V. I. GERSHOVICH, *Family of algorithms for solving convex programming problems*, Cybernetics, 15 (1979), pp. 502–508.

[102] M. SNIEDOVICH, *A new look at fractional programming*, Journal of Optimization Theory and Applications, 54 (1987), pp. 113–120.

[103] ——, *Fractional programming revisited*, European Journal of Operational Research, 33 (1988), pp. 334–341.

[104] ——, *Dynamic Programming*, no. 154 in Pure and applied mathematics, Marcel Dekker, New York, 1992.

[105] P. M. Vaidya, *A new algorithm for minimizing convex functions over convex sets*, tech. rep., AT&T Bell Laboratories, Murray Hill, New Jersey, 1989. To appear in Mathematical Programming.

[106] D. E. Varberg and A. W. Roberts, *Convex Functions*, Academic Press, New York, 1973.

[107] G. R. Walsh, *Methods of Optimization*, Wiley, London, 1975.

[108] J. E. Ward and R. E. Wendell, *A new norm for measuring distance which yields linear location problems*, Operations Research, 28 (1980), pp. 836–844.

[109] ——, *Using block norms for location modeling*, Operations Research, 33 (1985), pp. 1074–1090.

[110] R. E. Wendell and A. P. Hurther Jr., *Location theory, dominance and convexity*, Operations Research, 21 (1973), pp. 314–320.

[111] C. Witzgall, *On convex metrics*, Journal of Research of the National Bureau of Standards (section B), 69B (1965), pp. 175–177.

[112] G. R. Wood, *The bisection method in higher dimensions*, Mathematical Programming, 55 (1992), pp. 319–338.

[113] Y. Ye, *Complexity analysis of the analytic center cutting plane method that uses multiple cuts*, tech. rep., Department of Management Sciences, The University of Iowa, Iowa City, Iowa 52242, USA, 1994.

[114] D. B. Yudin and A. S. Nemirovsky, *Evaluation of the informational complexity of mathematical programming problems*, Matekon, 13 (1976), pp. 3–25.

[115] ——, *Informational complexity and efficient methods for the solution of convex extremal problems*, Matekon, 13 (1977), pp. 25–45.

[116] Y. Zabotin, *Methods of minimizing quasi-convex functionals*, PhD thesis, Kazan University, Kazan, Russia, 1977. in Russian.

[117] Y. Zabotin, A. Korablev, and R. Khabibullin, *On minimizing quasi-convex functionals*, Izvestiya VUZ. Matematika, (1972). in Russian.

[118] ——, *On the conditions of extremum of quasi-convex functional on convex set*, Kibernetika, (1973). in Russian.

[119] Y. ZABOTIN AND M. KREININ, *On constructing non-relaxation sub-gradient methods for convex programming*, Izvestiya VUZ. Matematika, (1983). in Russian.

[120] S. ZHANG, *Stochastic Queue Location Problems*, no. 14 in Tinbergen Institute Research Series, Thesis Publishers, Amsterdam, The Netherlands, 1991.

Index

MIX
Papier aus verantwortungsvollen Quellen
Paper from responsible sources
FSC® C105338

If you have any concerns about our products,
you can contact us on
ProductSafety@springernature.com

In case Publisher is established outside the EU,
the EU authorized representative is:
Springer Nature Customer Service Center GmbH
Europaplatz 3, 69115 Heidelberg, Germany

Printed by Libri Plureos GmbH
in Hamburg, Germany